D1117435

Theory and Application of Mathematical Programming

THEORY AND APPLICATION OF MATHEMATICAL PROGRAMMING

G. MITRA

Department of Statistics and Operational Research,
Brunel University, Uxbridge, United Kingdom

1976

ACADEMIC PRESS

London · New York · San Francisco

A Subsidiary of Harcourt Brace Jovanovich, Publishers

ACADEMIC PRESS INC. (LONDON) LTD.
24/28 Oval Road,
London NW1

United States Edition published by
ACADEMIC PRESS INC.
111 Fifth Avenue
New York, New York 10003

Library of Congress Catalog Card Number: 76-016981

ISBN: 0-12-500 450-8

Printed in Great Britain by Page Bros (Norwich) Ltd, Norwich

Preface

Mathematical Programming is going to occupy a central position in Applied Mathematics, comparable to that enjoyed by Differential Equations until a few decades ago. It may sound crass and prophetic, but this comparison is in many ways apt as both these topics have been developed to model physical situations, and both are used to obtain quantitative answers to questions concerning such situations.

The study of Mathematical Programming stems from two main motivations: one the identification of the problems which can be modelled by this technique and the other the development of theory and, more pertinently, solution techniques which may be applied to obtain quantitative answers. There is no dearth of text books on this topic, and it should indeed lift the heart of every proponent of Mathematical Programming that the entire field is in a state of growth. I do not find myself hard pressed to justify writing another text book. I admit being a good Hindu, hence always willing to accept a new point of view: and never too shy to present my own. It is the zeal for the latter that has prompted me to write this book. In many ways the book is an outgrowth of the lecture notes I prepared when I was a visiting lecturer at the Queen Mary College, University of London. In the last four years I have used them to teach the final year mathematics honours students at Brunel University. A total of fifty to sixty hours of lectures adequately covers the material in the book.

It is difficult to strike a balance between the theory and practice of Mathematical Programming. I have tried to achieve this: of course the relative weight for each is subjectively suggested by me. The use of keywords I consider is beneficial and is meant to direct the reader's attention to the pertinent concepts and notions. The first chapter of the book is meant to motivate the readers by presenting situations where Mathematical Programming problems arise and how these may be mathematically stated. The first part of the book comprising Chapters 2 to 6 covers the oft trodden topics of Linear Programming. By way of uniqueness I may claim that in Chapter 2 all the three alternative ways of representing the Linear Programming tableau, viz. Contracted Tableau, Extended Tableau, Tucker-Beale Form are presented. Up-to-date versions of Phase I method and Revised Simplex Method are provided in Chapter 3 and Chapter 4. Chapter 5 is intended to make the reader aware of the importance of the computational aspects of Linear Programming; indeed this has open horizons. Duality and Post Optimal Analysis are discussed in Chapter 6.

The second part of the book comprises non-linear programming; here only the computationally important Non-Linear Programming Algorithms and the problems to which these apply have been presented. An attempt is made to classify Integer Programs and Mixed Integer Programs. The techniques of Lagrange Multiplier and Kuhn–Tucker Multiplier are discussed; an algorithm for solving the Convex Quadratic Programming problem (by satisfying the Kuhn–Tucker conditions) is also outlined. In the last Chapter the algebra and the combinatorics of the complementary pivot theory is discussed: in many ways this is the most elegant theoretical development in recent times which has contributed towards unifying the theory of Mathematical Programming. The Appendix II and Appendix III are in many ways unique, for I am not aware of any other text book that discusses the method of use of an up-to-date large scale Mathematical Programming system, and a Mathematical Programming modelling language.

I am aware that the whole area of Network flow and the Transportation problem has been omitted; I do not wish to imply that these topics are not important. However, the advent of special algorithms to solve problems with Generalized Upper Bound (GUB) structure has shown in recent years that such problems are better analysed under GUB within the frame work of a Linear Programming Approach. In making this controversial statement I have the tacit support of more than one authority in this field.

I am particularly indebted to Professor K. Wolfenden, and Professor E. M. L. Beale. The former had introduced me to this field and the latter in his teachings and many discussions had explained to me and others who worked for him aspects of Mathematical Programming which could have otherwise remained grey and unilluminated. It is difficult to thank individually the devoted students who have helped in correcting many mistakes in the earlier versions of the manuscript. To correct the mistakes which still remain I look forward to the co-operation of the critical readers.

G. Mitra
August 1975

Contents

CHAPTER 1

What is Mathematical Programming?

KEYWORDS: Linear Programming, Nonlinear Programming. Mathematical Programming, Dynamic Programming, Integer Programming.

1.1 INTRODUCTION

Optimization or Programming problems are primarily concerned with the efficient use of limited resources; efficient use implying the best way of meeting a specified objective. From a mathematical standpoint solving one of these problems is equivalent to calculating the extremum of a function, the function in question having a domain (often called the feasible domain) over which it is defined. Many celebrated mathematicians had considered the methods of obtaining extrema of functions by analytic means [28]; such techniques are in general referred to as classical techniques.

About two decades ago, however, with the overwhelming motivation to obtain numerical answers to optimum planning problems posed quantitatively, G. B. Dantzig [15] in the year 1947 first proposed the "Simplex Method" whereby a linear form could be maximised or minimised subject to linear equalities or inequalities. Such problems have come to be known as "Linear Programs". Although this is not directly related to computer programs the development of linear programming has gone on hand in hand with the development of the computer.

For problems in which the objective function is nonlinear and/or the constraints are made up of nonlinear functions, the term "Nonlinear Programming" has been used. In order to be more general the word "Mathematical Programming" has been given currency and has become an all embracing term which includes both Linear and Nonlinear Programming. For clarity it is perhaps worth noting that in this context the word "programming", due to some freak of semantics, refers both to problems and their solution procedures.

Many time-dependent problems often go by the name of multistage problems, and these exhibit a sequential structure which permits solution by the application of an "inductive principle". Richard Bellman [7] in Rand Corporation was the first to see the importance of this principle which he termed the "principle of optimality"; the method of analysis and the solution technique developed by him has come to be known as "Dynamic Programming" [7], [27].

More recently [21] the methods of unconstrained optimization of functions have been extended to solve optimisation problems with constraints. The problems which may be solved by these techniques are often very different from the problems solved by mathematical programming techniques which are extensions of simplex type methods. Linear programming has been studied extensively from just before the 1950's to the present day and there are many contributors and textbooks in this field. The textbooks with which the author is most familiar are [1], [2], [15], [26], [45], [52]. There are two areas of extension viz: Networkflow problems [22] and Integer programming problems [2], [15], [27], [45] which have proven to be the natural and most important outgrowth of linear programming studies. Some classes of combinatorial and discrete problems such as the assignment problem, the shortest route problem and the transportation problem, may be posed as linear flow problems. Likewise, there are large classes of non-linear non-convex discrete, combinatorial problems which may be reduced to integer programming problems. An "Inter Program" may be defined as a linear program with the added condition that some or all the variables of the problem must be integer valued. In the rest of this chapter examples of problems which arise in different contexts and their mathematical formulation are presented.

1.2 MATHEMATICAL FORMULATION OF OPTIMIZATION OR PROGRAMMING PROBLEMS

(i) *Transportation problem*

Let there be m sources each having a single commodity (say fuel) in quantities a_i ($i = 1, 2, \ldots, m$) and let there be n sinks each having demands b_j ($j = 1, 2, \ldots, n$) for this commodity. The transportation problem is that of finding an optimal plan of transporting this commodity from the sources to the sinks. Let x_{ij} be the number of units of the product transported from source i to destination j via the route (i, j) at cost C_{ij} per unit. Assume that the cost relationship is linear, i.e. the cost of transporting x_{ij} units over route (i, j) is

	Destinations (n) 1	2	3	4	5	Availabilities (a_i)
Sources (m) 1	C_{11}, x_{11}	C_{12}, x_{12}	C_{13}, x_{13}	C_{14}, x_{14}	C_{15}, x_{15}	a_1
2	C_{21}, x_{21}	C_{22}, x_{22}	C_{23}, x_{23}	C_{24}, x_{24}	C_{25}, x_{25}	a_2
3	C_{31}, x_{31}	C_{32}, x_{32}	C_{33}, x_{33}	C_{34}, x_{34}	C_{35}, x_{35}	a_3
Requirements (b_j)	b_1	b_2	b_3	b_4	b_5	

$C_{ij} \cdot x_{ij}$. One has to distribute the availabilities $\sum_{i=1}^{m} a_i$ over the requirements

$$\sum_{j=1}^{n} b_j \text{ where } \left(\sum_{i=1}^{m} a_i = \sum_{j=1}^{n} b_j \right)$$

so as to minimise the total transport cost:

$$\text{Minimise} \qquad \sum_{i=1}^{m} \sum_{j=1}^{n} C_{ij} \cdot x_{ij}, \tag{1.1}$$

subject to $x_{ij} \geqslant 0$ (one way flow), (1.2)

and

$$\left.\begin{aligned} \sum_{j=1}^{n} x_{ij} &= a_i, \text{ (m supply equations)} \\ \sum_{i=1}^{m} x_{ij} &= b_j, \text{ (n demand-equations)} \end{aligned}\right\} \begin{array}{l} m + n \text{ equation.} \\ \text{in } m \times n \text{ variables} \end{array} \tag{1.3}$$

The problem admits of integer solution if a_i, b_j are non-negative integers.

(ii) *Production planning problem*

A factory has m assembly lines each of which can produce any one of n different types of product. The ith assembly line ($i = 1, 2, \ldots, m$) can produce one unit of the jth product ($j = 1, 2, \ldots, n$) in a time t_{ij} at a cost c_{ij}, but cannot operate for more than a time b_i per week. The factory is required to produce at least a_j units of the product j per week; let x_{ij} (the units of the jth product on the ith assembly line) be the variable in which the problem is defined. One has to determine the production programme which fulfils the output requirement at the minimum cost:

$$\text{Minimize} \qquad \sum_{i=1}^{m} \sum_{j=1}^{n} c_{ij} \, x_{ij}, \tag{1.4}$$

subject to

$$x_{ij} \geqslant 0 \qquad (m \times n \text{ variables})$$

$$\sum_{i=1}^{m} x_{ij} \geqslant a_j \qquad (n\text{-inequalities}) \tag{1.5}$$

$$\sum_{j=1}^{n} x_{ij} t_{ij} \leqslant b_i \qquad (m\text{-inequalities}).$$

(iii) *Choice of power plants in electricity generation system*

An electricity supply authority has the following commitments to meet in its next planning period.

Firm power: Average rate of consumption A (in MW) during the heavy loading in winter days.

Peak power: Average rate of consumption P (in MW) during the four peak hours of heavy loading in winter days.

Annual consumption: Total energy requirement during a year E (MWh) of the loads being supplied by the board.

Let the following ranges of power plants be available in the market: the characteristics of the power plants are shown in Table 1.1.

TABLE 1.1

Index	Type of power plant	Peaking factor (p_i)	Annual production (e_i) (MWh)	Capital cost (d_i) $\left(\frac{\text{sterling}}{\text{MW}}\right)$	Objective cost coefficient (c_i) $\left(\frac{\text{sterling}}{\text{MW}}\right)$	Technological bound (b_i) (MW)
1	Coal fired steam	p_1	e_1	d_1	c_1	b_1
2	Gas fired steam	p_2	e_2	d_2	c_2	b_2
3	Gas turbine	p_3	e_3	d_3	c_3	b_3
4	Hydro (I)	p_4	e_4	d_4	c_4	b_4
5	Hydro (II)	p_5	e_5	d_5	c_5	b_5
6	Tidal power	p_6	e_6	d_6	c_6	b_6
7	Nuclear	p_7	e_7	d_7	c_7	b_7

p_i: The ratio, peak output/maximum continuous rating for the ith type power plant.
e_i: The Annual production capacity in MWh of the ith type power plant of unit rating.
d_i: Capital cost £/MW for the ith type power plant of unit rating.
c_i: The objective cost coefficient of the ith type power plant as deduced from capital cost, maintenance cost, capitalization factor and running cost. It has the unit of £/MW.
b_i: The upper bound on the MW capacity of the ith type power plant that can be installed.

The supply authority has at its disposal a budget of D pounds.

The supply authority has therefore to choose the amount of plant capacity x_i of type i in MW which must be installed to satisfy the constraints:

$$\left.\begin{array}{ll} x_1 + x_2 + \ldots + x_7 \geqslant A & \text{Firm power commitment} \\ p_1x_1 + p_2x_2 + \ldots + p_7x_7 \geqslant P & \text{Peak power commitment} \\ e_1x_1 + e_2x_2 + \ldots e_7x_7 \geqslant E & \text{Energy commitment} \\ d_1x_1 + d_2x_2 + \ldots + d_7x_7 \leqslant D & \text{Budget constraint} \\ \left.\begin{array}{l} x_i \leqslant b_i \\ x_i \geqslant 0 \end{array}\right\} i = 1, 2, \ldots, 7 & \begin{array}{l}\text{Bounding and non-} \\ \text{negativity constraints}\end{array} \end{array}\right\} \quad (1.6)$$

and find the minimum of the objective function,

$$\text{Minimize } c_1x_1 + c_2x_2 + \ldots + c_7x_7. \quad (1.7)$$

(iv) *Optimum redundancy of multistage system* [44]

A series parallel system of k stages is such that the system fails if one particular stage fails. Each stage j can have n_j components in parallel and the stage j functions if at least one of the components of the stage j functions. The independent probability of a unit of the jth type functioning is p_j and failing is $q_j = 1 - p_j$. The probability of failure of the jth stage with n_j components is $q_j^{n_j}$. One has to find the maximum of the system reliability function $R(n)$— the probability of the whole system functioning,

$$R(n_1, n_2, n_3, \ldots, n_k) = \prod_{j=1}^{j=k} (1 - q_j^{n_j}) \quad (1.8)$$

which must satisfy the overall weight constraint

$$w_1n_1 + w_2n_2 + \ldots + w_kn_k \leqslant W$$

and the cost constraint, (1.9)

$$c_1n_1 + c_2n_2 + \ldots + c_kn_k \leqslant C$$

further $n_j \geqslant 1, j = 1, 2, \ldots, k$ must be positive integers

where w_j and c_j are the weight and cost coefficients of the jth stage and W, C are the allowable weight and cost respectively.

(v) *Tower spotting problem* [37]

Given the survey data of a transmission line route and the choice of available towers of standard, i.e., suspension type and of angle towers, it is required to choose and site the towers in such a way that the overall cost of running the line from one end of the route to the other is minimum.

Assume that the following information is available,

A_{hot} the parameter of the parabola in which the line is assumed to hang at 122°F, the statutory design temperature,
x the horizontal distance from the beginning of the line section,
$s_1(x)$ the maximum single span limit at x,
s_2 the maximum double span limit,
X_{test} an ordered set of test sites,
X_{tower} an ordered set of tower sites,
$x_{1,i}$ the distance from x to the lowest point of the span, (x_i, x_{i+1}),
$x_{-1,i}$ the distance from x to the lowest point of the span, (x_{i-1}, x_i),
$z(x)$ statutory clearance from ground or any other object at x,
H_{tower} an ordered set of available tower heights,
$v(x)$ the elevation above sea level of the lowest conductor of the line at x
$u(x)$ the elevation above sea level of the ground or any offending structure at distance x.
$c(x, h)$ the cost of erecting a tower of height h at the site x.

The problem can now be formulated as follows. Choose an ordered set of tower sites

$$X_{tower} = \{x_i | i = 0, 1, 2, \ldots, M, \text{ and } x_{i-1} < x_i\} \tag{1.10}$$

from an ordered set of test sites

$$X_{test} = \{X_j | j = 0, 1, 2, \ldots, n, \text{ and } X_{j-1} < X_j\} \tag{1.11}$$

where $X_{tower} \subset X_{test}$, in such a way as to minimise the cost function

$$\sum_{i=1}^{m} c(x_i, h_i). \tag{1.12}$$

The tower height h_i at x_i is selected from

$$H_{tower} = \{H_k | k = 1, 2, \ldots, \mathbf{q} \text{ and } H_{k-1} < H_k\}.$$

The end points of the section may be more explicitly specified by noting that $x_0 = X_0$ and $x_m = X_n$. The minimization of this problem just stated has to be carried out subject to the constraints:

(a) a maximum single span constraint

$$x_i - x_{i-1} \leqslant s_1(x_i), \qquad i = 1, 2, \ldots, M, \tag{1.13}$$

(b) a maximum double span constraint

$$x_{i+1} - x_{i-1} \leqslant s_2, \qquad i = 1, 2, \ldots, M - 1, \tag{1.14}$$

(c) an uplift constraint which sets a limit on the possible angle of swing of the line under transverse wind loading

$$x_{1,i} + x_{-1,i} \geqslant W(x_{i+1} - x_{i-1}), \qquad i = 1, 2, \ldots, M-1, \qquad (1.15)$$

W being a parameter of the line constants,

(d) a statutory clearance constraint to conform with the British Standard Specifications

$$v(x) - u(x) > Z(x), \qquad (1.16)$$

where the elevation of the line conductor is given by

$$v(x) = A_{\text{hot}}(x - x_i)(x - x_{i+1}) + \left(\frac{v_{i+1} - v_i}{x_{i+1} - x_i}\right)(x - x_i) + v_i \qquad (1.17)$$

and $v_i = v(x_i) \ldots$ etc.

These are a few examples of mathematical programming problems.

CHAPTER 2

Equivalent Linear Programming Problems and The Simplex Method

KEYWORDS: Canonical Form, Slack Variables, Structural Variables, Free Variables, General Form, Standard Form, Mixed Form, Basis, Feasible Solution, Basic Solution, Basic Feasible Solution, Pivotal Transformation, Reduced Cost Coefficient, Shadow Price.

2.1 STATEMENT OF THE LINEAR PROGRAMMING PROBLEM

The linear programming problem may be stated in the "Canonical Form"

$$\text{Maximize } f(x) = c_1x_1 + c_2x_2 + \ldots + c_nx_n,$$

a linear function (in n-variables) subject to the linear constraints,

$$\begin{array}{l} a_{11}x_1 + a_{12}x_2 + \ldots + z_{1n}x_n \leqslant b_1 \\ a_{21}x_1 + a_{22}x_2 + \ldots + a_{2n}x_n \leqslant b_2 \\ \quad\vdots \qquad\quad \vdots \qquad\qquad\quad \vdots \qquad \vdots \\ a_{m1}x_1 + a_{m2}x_2 + \ldots + a_{mn}x_n \leqslant b_m. \end{array} \tag{2.1}$$

The variables are further required to be non-negative, i.e.

$$x_1, x_2, \ldots, x_n \geqslant 0.$$

In matrix notation this can be expressed as

$$\begin{array}{ll} \text{maximize} & f(x) = c'x \\ \text{subject to} & Ax \leqslant b \\ \text{and} & \quad x \geqslant 0. \end{array} \tag{2.2}$$

To pose the problem (2.1) in an equivalent form, where the inequalities are replaced by equalities, m "Slack Variables"† are introduced x_{n+1}, x_{n+m}, as distinct from the n "Structural Variables"‡ in which the problem is defined. The problem is now restated with the equality relations,

$$\begin{aligned} &\text{maximize } f(x) = c_1x_1 + c_2x_2 + \ldots + c_nx_n \\ &\text{subject to } a_{11}x_1 + a_{12}x_2 + \ldots + a_{1n}x_n + x_{n+1} = b_1 \\ &\qquad a_{21}x_1 + a_{22}x_2 + \ldots + a_{2n}x_n + x_{n+2} = b_2 \\ &\qquad \vdots \\ &\qquad a_{m1}x_1 + a_{m2}x_2 + \ldots + a_{mn}x_n + x_{m+n} = b_m, \end{aligned} \tag{2.3}$$

and

$$x_1, x_2, \ldots x_{n+m} \geqslant 0.$$

In matrix notation this is expressed as

$$\begin{aligned} &\text{maximize } f(x) = c'x \\ &\text{subject to } Ax + Ix_{sL} = b \end{aligned} \tag{2.4}$$

where $x = (x_1, x_2, \ldots, x_n) \geqslant 0$, and $x_{sL} = (x_{n+1}, x_{n+2}, \ldots, x_{n+m}) \geqslant 0$, and A is an $m \times n$ matrix, I is an $m \times m$ identity matrix, x is an n-vector and x_{sL} and b are m-vectors.

2.2 EQUIVALENT FORMS OF LINEAR PROGRAMS

There are three further forms which are often referred to in the literature. These are "General Form", "Standard Form" and "Mixed Form".

(a) General form

$$\text{Maximize } \sum_{j=1}^{n} c_jx_j$$

$$\text{subject to } \sum_{j=1}^{n} a_{ij}x_j \leqslant b_i, \qquad i = 1, 2, \ldots, p$$

$$\sum_{j=1}^{n} a_{ij}x_j = b_i, \qquad i = p+1, p+2, \ldots, m \tag{2.5}$$

and

$$x_j \geqslant 0 \quad \text{for} \quad j = 1, 2, \ldots, q$$

$$-\alpha \leqslant x_j \leqslant \alpha \quad \text{for} \quad j = q+1, q+2, \ldots, n.$$

† Also called "logical variables".
‡ Also referred to as "natural variables".

The second set of variables are often referred to as "Free variables" in the literature.

(*b*) *Standard form*

$$\text{Maximize } f(x) \sum_{j=1}^{n} c_j x_j,$$

$$\text{subject to} \quad \sum_{j=1}^{n} a_{ij} x_j = b_i, \quad i = 1, 2, \ldots, m, \tag{2.6}$$

and

$$x_j \geqslant 0, \quad j = 1, 2, \ldots, n.$$

(*c*) *Mixed form*

$$\text{Maximize } f(x) = \sum_{j=1}^{n} c_j x_j,$$

$$\text{subject to} \quad \sum_{j=1}^{n} a_{ij} x_j = b_i \quad i = 1, 2, \ldots, p \tag{2.7}$$

$$\sum_{j=1}^{n} a_{ij} x_j \leqslant b_i \quad i = p+1, p+2, \ldots, m$$

and

$$x_j \geqslant 0, \quad j = 1, 2, \ldots, n.$$

It may be shown [45] by one or more of the following operations that all these four forms, "Canonical", "General", "Standard", and "Mixed" are equivalent to one another.

Operation 1. A free variable x_j may be replaced by

$$x_j = -y_j + x'_j \tag{2.8}$$

where an appropriate combination of non-negative values of y_j and x'_j may represent a free value of x_j. An alternative way of treating free variables is given in [45]; computer codes use yet another technique to deal with these.

Operation 2. An equation

$$a_{i1}x_1 + a_{i2}x_2 + \ldots + a_{in}x_n = b_i \tag{2.9}$$

may be replaced by the two inequalities,

$$a_{i1}x_1 + a_{i2}x_2 + \ldots + a_{in}x_n \leqslant b_i \tag{2.10}$$

and

$$-a_{i1}x_1 - a_{i2}x_2 - \ldots - a_{in}x_n \leqslant -b_i.$$

Operation 3. An inequality

$$a_{i1}x_1 + a_{i2}x_2 + \ldots + a_{in}x_n \leqslant b_i \tag{2.11}$$

may be replaced by the equality

$$a_{i1}x_1 + a_{i2}x_2 + \ldots + a_{in}x_n + x_{n+i} = b_i \tag{2.12}$$

and the non-negativity constraint $x_{n+i} \geqslant 0$.

Using these operations the following transformations from one form to another are possible.

$$\text{General form} \xrightarrow{\text{(Operation 1)}} \text{Mixed form,}$$

$$\text{Mixed and standard forms} \xrightarrow{\text{(Operation 2)}} \text{Canonical form,}$$

$$\text{Mixed and canonical forms} \xrightarrow{\text{(Operation 3)}} \text{Standard form.}$$

2.3 THE SIMPLEX METHOD EXPLAINED BY AN EXAMPLE

Consider the problem,

$$\begin{aligned} &\text{Maximize } f(x) = 5x_1 + 4x_2 \\ &\text{subject to} \quad 2x_1 + 4x_2 \leqslant 1000 \\ &\qquad\qquad\quad 2x_1 + x_2 \leqslant 400 \\ &\text{and } x_1, x_2 \geqslant 0. \end{aligned} \tag{2.13}$$

Introducing slack variables x_3, x_4 ($x_3, x_4 \geqslant 0$) one obtains the equivalent problem

$$\begin{aligned} &\text{Maximize } f(x) = 5x_1 + 4x_2 \\ &\text{subject to } 2x_1 + 4x_2 + x_3 = 1000 \\ &\qquad\qquad\quad 2x_1 + x_2 + x_4 = 400. \end{aligned} \tag{2.14}$$

Figure 2.1 provides a graphical representation of this problem; note that $x_1 = 0$, $x_2 = 0$, $x_3 = 0$, $x_4 = 0$ represent lines/hyperplanes which define a closed convex† region. Any solution for which the x variables are non-negative and the equations are satisfied is called feasible.

† Convexity is a mathematical property relating to functions and solution spaces; for a preliminary mathematical treatment see [45].

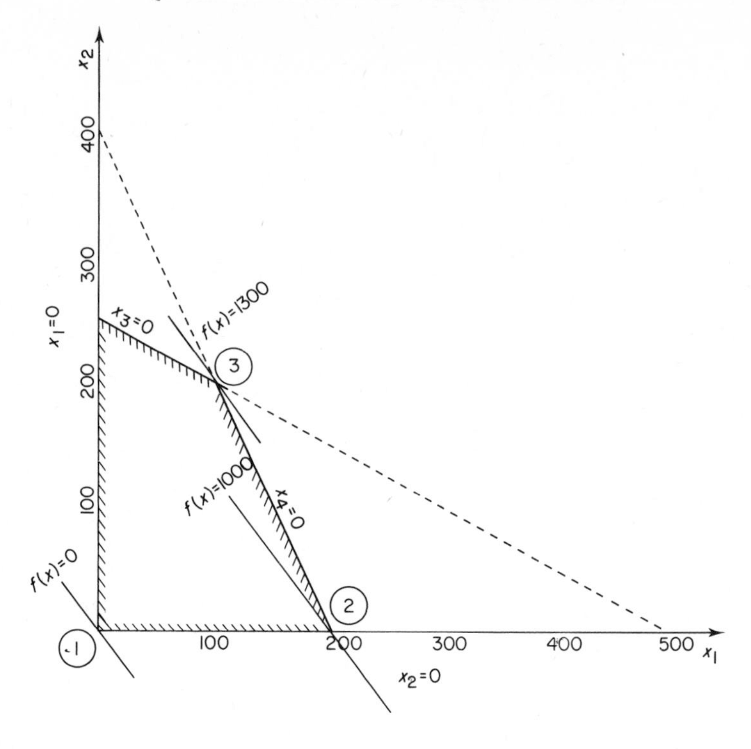

FIG. 2.1.

Rewriting the equation set (2.14) as

$$\begin{aligned} f(x) &= 0 + 5x_1 + 4x_2 \\ x_3 &= 1000 - 2x_1 - 4x_2 \\ x_4 &= 400 - 2x_1 - x_2 \end{aligned} \tag{2.15}$$

it is observed that an increase in the value of x_1 (positive cost coefficient) increases the function value. For all practical purposes the columns of the other variables in this problem may be forgotten; the range of the change in x_1 (not considering the variable x_2) can be:

$$\begin{aligned} f(x) &= 0 + 5x_1 + 4(0) \\ x_3 &= 1000 - 2x_1 - 4(0)\text{: range of } x_1 \text{ is } 1000/2 = 500 \\ x_4 &= 400 - 2x_1 - (0)\text{: range of } x_1 \text{ is } 400/2 = 200. \end{aligned} \tag{2.16}$$

pivotal element (points to $2x_1$ in the x_4 equation)

In order to remain feasible i.e., right hand side numerical values non-negative, the lowest of the range is chosen, and x_1 is rewritten in terms of x_4 (see 2.15),

$$x_1 = 200 - x_4/2 - x_2/2. \tag{2.17}$$

Substituting (2.17) in the other two equations of (2.15) the new equation set (2.18) is obtained:

$$\begin{aligned} f(x) &= 1000 - 5/2x_4 + 3/2x_2 \\ x_3 &= 600 + x_4 - 3x_2 \\ x_1 &= 200 - x_4/2 - x_2/2. \end{aligned} \tag{2.18}$$

This represents a change of state. The step whereby the equation set (2.15) is transformed into the equation set (2.18) is called a "Pivotal Transformation" and the chosen (dividing) element 2 the pivotal element.

The function $f(x)$ can still be increased by moving in the direction of x_2. This time the ranges of x_2 are $600/3 = 200$ and $200/(\frac{1}{2}) = 400$, therefore the second equation in (2.18) must be rewritten to maintain feasibility,

$$x_2 = 200 + x_4/3 - x_3/3. \tag{2.19}$$

Substituting this in the other equations of (2.18) the transformed set of equations becomes

$$\begin{aligned} f(x) &= 1300 - 2x_4 - x_3/2, \\ x_2 &= 200 + x_4/3 - x_3/3 \\ x_1 &= 100 - 2/3x_4 + x_3/6 \end{aligned} \tag{2.20}$$

Inspecting the cost coefficients of the equation set (2.20) it is observed that $f(x)$ cannot be increased any further. Setting $x_4 = 0, x_3 = 0$, the optimum solution $f(x) = 1300$, $x_1 = 100$, $x_2 = 200$ is obtained.

2.4 SOME TERMINOLOGY AND CONCEPTS

At this stage the reader should familiarize himself with the problem of solving a set of linear equations with n variables and m constraints where $m \leqslant n$ and the variables are required to be non-negative. In the Appendix I the relevant concepts are expounded.

Refer to the equation set (2.3) and rewrite this as

$$\begin{aligned}
&a_{11}x_1 + a_{12}x_2 \ldots + a_{1n}x_n + a_{1n+1}x_{n+1} + a_{1n+m}x_{n+m} = b_1 \\
&a_{21}x_1 + a_{22}x_2 \quad + a_{2n}x_n + a_{2n+1}x_{n+1} + a_{2n+m}x_{n+m} = b_2 \\
&\vdots \\
&a_{m1}x_1 + a_{m2}x_2 \quad + a_{mn}x_n + a_{mn+1}x_{n+1} + a_{mn+m}x_{n+m} = b_m,
\end{aligned} \tag{2.21}$$

where note $a_{1n+1} a_{2n+2}, \ldots, a_{mn+m} = 1$, and $a_{i,n+j} = 0$ for $i = 1, \ldots, m$, $j = 1, \ldots, m$ and $i \neq j$. Also consider another equivalent form of the set of equations obtained after m number of transformations of the type illustrated in the last section,

$$\begin{aligned}
x_1 + \quad & + \bar{a}_{1m+1}x_{m+1} + \ldots + \bar{a}_{1m+n}x_{m+n} = \beta_1 \\
x_2 \quad & + \bar{a}_{2m+1}x_{m+1} + \ldots + \bar{a}_{2m+n}x_{m+n} = \beta_2 \\
\ddots \quad & \\
x_m & + \bar{a}_{mm+1}x_{m+1} + \ldots + \bar{a}_{mn+n}x_{m+n} = \beta_m.
\end{aligned} \tag{2.22}$$

Without any loss of generality it may be assumed that the matrices B_1 and B_2

$$B_1 = \begin{bmatrix} a_{1n+1} & & a_{1n+m} \\ & & \\ a_{mn+1} & & a_{mn+m} \end{bmatrix} = \begin{bmatrix} 1 & & \\ & 1 & \\ & & 1 \end{bmatrix} \text{ and } B_2 = \begin{bmatrix} a_{11} & a_{12} & a_{1m} \\ & & \\ a_{m1} & a_{m2} & a_{mm} \end{bmatrix}$$

are any two square submatrices of the matrix of the coefficients of the set of equations in (2.21).

If $\det(B_1) \neq 0$, and $\det(B_2) \neq 0$ i.e., B_1, B_2 are non-singular then the two sets of equations are solvable as

$$\begin{bmatrix} & & \\ & B_1 & \\ & & \end{bmatrix} \cdot \begin{bmatrix} x_{n+1} \\ x_{n+2} \\ \vdots \\ x_{n+m} \end{bmatrix} = \begin{bmatrix} b_1 \\ b_2 \\ \vdots \\ b_m \end{bmatrix} \quad \text{or} \quad \begin{aligned} x_{n+1} &= b_1 \\ x_{n+2} &= b_2 \\ &\vdots \\ x_{n+m} &= b_m \end{aligned} \tag{2.23}$$

and

$$\begin{bmatrix} & & \\ & B_2 & \\ & & \end{bmatrix} \cdot \begin{bmatrix} x_1 \\ x_2 \\ \\ x_m \end{bmatrix} = \begin{bmatrix} b_1 \\ \cdot \\ \\ b_m \end{bmatrix} \quad \text{or} \quad \begin{aligned} x_1 &= \beta_1 \\ x_2 &= \beta_2 \\ &\vdots \\ x_m &= \beta_m, \end{aligned} \tag{2.24}$$

The solutions in (2.23) and (2.24) are contained in the two equivalent representations (2.21), and (2.22). The two matrices B_1, B_2 from which the solutions are derived are called the "Basis Matrix". The variables corresponding to the appropriate columns of the basis matrix are called "Basic Variables". From (2.21) and (2.22) it is observed that each of the basic variables can be expressed as the sum or difference of a solution value and a set of other variables multiplied by constant coefficients. Such other variables as x_1, $x_2, \ldots, x_n$ in (2.21), and $x_{m+1}, x_{m+2} \ldots x_{n+m}$ in (2.22) are called the "non-basic variables". A solution involving a set of m basic variables is called a "Basic Solution". Such solutions as in (2.23) and (2.24) become identical to that in (2.21) and (2.22) if the set of non-basic variables are set to zero. A solution to the set of equations is said to be a "Feasible Solution" if the non-negativity requirement of the linear programming constraints are satisfied. A basic solution which is also feasible is called a "Basic Feasible" solution.

In the example of the last section

$$\left.\begin{array}{ll} x_3 = 1000, & x_4 = 400 \\ x_1 = 0, & x_2 = 0 \end{array}\right\}, \text{ as in (2.15)}$$

$$\left.\begin{array}{ll} x_3 = 600, & x_1 = 200 \\ x_4 = 0, & x_2 = 0 \end{array}\right\}, \text{ as in (2.18)}$$

and

$$\left.\begin{array}{ll} x_2 = 200, & x_1 = 100 \\ x_4 = 0, & x_3 = 0 \end{array}\right\}, \text{ as in (2.20)}$$

correspond to three sets of basic solutions which are also feasible. In linear programming one is most concerned with basic feasible solutions since a fundamental theorem of linear programming states that the optimum solution to a linear programming problem if it exists must be a basic feasible solution.

EXERCISES

2.1 In the example (2.13) if one sets $x_1 = 100$, $x_2 = 100$, $x_3 = 400$, $x_4 = 100$, will this solution be feasible? Will this solution be basic? Observe that every feasible solution is not necessarily basic.

2.2 If x_1 and x_4 form the basis what will be the solution values? Are all the values of the variables non negative? Observe that every basic solution is not necessarily feasible.

2.5 REPRESENTATION OF LINEAR PROGRAMMING PROBLEMS IN THE FORM OF TABLEAUX AND THE PIVOTAL TRANSFORMATION RULES

The matrix coefficients of a linear programming problem for the purpose of hand computation and simple computer implementation are written in the forms of tables known as "Tableaux". A representation due to A. W. Tucker known as the "Contracted Tableau" or "Tucker-diagram" takes the following form. Rewrite the equation set (2.3) as:

$$\begin{aligned} f(x) &= 0 - c_1(-x_1) - c_2(-x_2) - \ldots - c_n(-x_n) \\ x_{n+1} &= b_1 + a_{11}(-x_1) + a_{12}(-x_2) + \ldots + a_{1n}(-x_n) \\ x_{n+2} &= b_2 + a_{21}(-x_1) + a_{22}(-x_2) + \ldots + a_{2n}(-x_n) \\ &\vdots \\ x_{n+m} &= b_m + a_{m1}(-x_1) + a_{m2}(-x_2) + \ldots + a_{mn}(-x_n). \end{aligned} \tag{2.25}$$

Tableau 2.1 contains only the matrix coefficients and provides a compact representation of the equation set (2.25). The variables are noted along the margin: this also serves to illustrate the dependence of the variables in the rows (to the left) on the variables in the columns (along the top) multiplied by the coefficients.

		$-x_1$	$-x_2$	$\ldots$	$-x_n$
$f(x)$	0	$-c_1$	$-c_2$	$\ldots$	$-c_n$
x_{n+1}	b_1	a_{11}	a_{12}	$\ldots$	a_{1n}
x_{n+2}	b_2	a_{21}	a_{22}	$\ldots$	a_{2n}
$\vdots$	$\vdots$	$\vdots$	$\vdots$		$\vdots$
x_{n+m}	b_m	a_{m1}	a_{m2}	$\ldots$	a_{mn}

TABLEAU 2.1

Consider the pth equation in (2.25) and rewrite it in terms of the qth variable then

$$x_{n+p} = b_p + a_{p1}(-x_1) + a_{p2}(-x_2) + \ldots + a_{pq}(-x_q) + \ldots + a_{pn}(-x_n)$$

is written as

$$x_q = \frac{b_p}{a_{pq}} + \frac{a_{pl}}{a_{pq}}(-x_1) + \frac{a_{p2}}{a_{pq}}(-x_2) + \ldots + \frac{1}{a_{pq}}(-x_{n+p}) + \ldots + \frac{a_{pn}}{a_{pq}}(-x_n) \tag{2.26}$$

provided $a_{pq} \neq 0$.

Substituting the value of x_q in the ith equation, the following relationship is obtained after transformation,

$$x_{n+i} = b_i + a_{i1}(-x_1) + a_{i2}(-x_2) - \ldots - a_{iq}\left\{\frac{b_p}{a_{pq}} + \frac{a_{p1}}{a_{pq}}(-x_1) + \ldots + \frac{a_{p2}}{a_{pq}}(-x_2) + \ldots + \frac{1}{a_{pq}}(-x_{n+p}) + \ldots + \frac{a_{pn}}{a_{pq}}(-x_n)\right\} \ldots a_{in}(-x_n)$$

which after rearranging becomes

$$x_{n+i} = \left(b_i - \frac{a_{iq}b_p}{a_{pq}}\right) + \left(a_{i1} - \frac{a_{p1}a_{iq}}{a_{pq}}\right)(-x_i) \ldots - \frac{a_{iq}}{a_{pq}}(-x_{n+p}) + \ldots\left(a_{in} - \frac{a_{pn}a_{iq}}{a_{pq}}\right)(-x_n). \tag{2.27}$$

In order to be general and also to relate the coefficients of the transformed equations to the entries of the contracted tableau let a typical element $\bar{a}_{ij}$ at a certain step of the tableau representation be considered. The rules of pivotal transformation (by which the subsequent tableau element $\bar{a}'_{ij}$ are obtained) are set out below.

Let p and q denote the indices of the pivot row and the pivot column and $\bar{a}_{pq} \neq 0$; then the transformed elements are given by the relation

$$\begin{aligned} \bar{a}'_{pq} &= 1/\bar{a}_{pq}, \\ \bar{a}'_{pj} &= \bar{a}_{pj}/\bar{a}_{pq}, \quad j = 0, 1, \ldots, n \quad \text{and} \quad j \neq q, \\ \bar{a}'_{iq} &= -\bar{a}_{iq}/\bar{a}_{pq}, \quad i = 0, 1, 2, \ldots, m \quad \text{and} \quad i \neq p, \\ \bar{a}'_{ij} &= \bar{a}_{ij} - \bar{a}_{iq}\bar{a}_{pj}/\bar{a}_{pq} \text{ for all } i, j \text{ except } i = p \text{ or } j = q. \end{aligned} \tag{2.28}$$

Note that the first row always contains the transformed cost coefficients $\bar{c}'_j = \bar{a}'_{0j}$ called the reduced cost coefficients and $\bar{\beta}'_i = \bar{a}'_{i0}$ the solution values are contained in the first column; the same transformation rules (2.28) apply.

2.6 STATEMENT OF THE SIMPLEX ALGORITHM

At this stage ignore the term "Dual" in the statement of the algorithm. Primal variables are the structural and slack variables in which a problem is stated (primal problem) and solved.

If the problem is primal (dual) feasible, that is

$$\bar{a}_{i0} = \beta_i \geqslant 0, \qquad i \geqslant 1 \qquad (\bar{a}_{0j} = \bar{c}_j \geqslant 0, j \geqslant 1),$$

choose a column q (row p) with first element $\bar{a}_{0q}(\bar{a}_{p0})$ negative. From among the positive (negative) elements in this column (row) select the one for which the ratio $\bar{a}_{i0}/\bar{a}_{iq}$ $(\bar{a}_{0j}/\bar{a}_{pj})$ attains its least absolute value. This is the pivot element $\bar{a}_{pq} \neq 0$. If no positive (negative) element exists in such a column (row) then the problem is unbounded (not feasible). The optimal solution is obtained when $(\bar{a}_{i0} = \beta_i \geqslant 0)$ primal feasibility and $(\bar{a}_{0j} = \bar{c}_j \geqslant 0)$ dual feasibility are attained together. There can be but three outcomes to a linear program:

(1) There exists an "optimum" feasible solution and the primal simplex method can find this.
(2) The solution may be "unbounded" i.e. the convex region is not bounded and one can indefinitely increase the function value and there exists no optimum. Again the primal simplex method detects this.
(3) The constraint set may be inconsistent i.e. "no feasible" solution may exist, dual simplex method or the modified primal simplex method can find this (see next chapter).

Note that in the primal simplex method,

(a) a column q is first chosen by the (first or most) negative reduced cost criterion,

$$\bar{a}_{0q} = \min_j \{\bar{a}_{0j} | \bar{a}_{0j} < 0 \text{ and } j = 1, 2, \ldots, n\}.$$

(b) A row p is then chosen by the minimum ratio criterion for strictly positive pivots,

$$\frac{\bar{a}_{p0}}{\bar{a}_{pq}} = \min_i \left\{ \frac{\bar{a}_{i0}}{\bar{a}_{iq}} \,|\, \bar{a}_{i0} \geqslant 0 \text{ and } \bar{a}_{iq} > 0; i = 1, 2, \ldots, m \right\}. \tag{2.30}$$

Thus starting with a feasible solution there can be only two exits in the algorithm: no negative reduced cost coefficients (all $a_{0j} \geqslant 0$) i.e. optimum, or no positive pivot in a column with negative reduced cost i.e., the problem is unbounded. The problem (2.13) is solved in the tableau form as shown below.

		$-x_1$	$-x_2$
$f(x)$	0	-5	-4
x_3	1000	2	4
x_4	400	(2)	1

TABLEAU 2.2

		$-x_4$	$-x_2$
$f(x)$	1000	$\frac{5}{2}$	$-\frac{3}{2}$
x_3	600	-1	(3)
x_1	200	$\frac{1}{2}$	$\frac{1}{2}$

TABLEAU 2.3

		$-x_4$	$-x_3$
$f(x)$	1300	2	$\frac{1}{2}$
x_2	200	$-\frac{1}{3}$	$\frac{1}{3}$
x_1	100	$\frac{2}{3}$	$-\frac{1}{6}$

TABLEAU 2.4

The transformed cost coefficients from tableau to tableau are called "reduced cost coefficients"; due to tradition in notation practitioners in the field call these D_j. In Tableau 2.2–4 is the cost coefficient of the variable x_2 and after pivotal transformation the new cost coefficient becomes $-3/2$ in Tableau 2.3; This is the D_j or the reduced cost coefficient. The slack variables did not have cost coefficients originally, but after transformation these have non-zero cost coefficients, i.e. the objective function becomes dependent on these. Reduced cost coefficients associated with the slack variables are called "shadow prices". Shadow prices can be interpreted as a measure of the change of the objective function value for unit change in the r.h.s. value of the associated constraint or the original slack variable value; provided no basis change was necessary to maintain optimality.

EXERCISE

2.3 In the worked out example the shadow price for the slack variable x_4 is $+2$ which means if the second constraint was increased to

$$2x_1 + x_2, \qquad \leqslant 500,$$

an increase of 100 in the r.h.s. value. The function value will become $1300 + 2 \times 100 = 1500$. Verify this by resolving the problem, with this new constraint.

2.7 EXTENDED TABLEAU AND TUCKER-BEALE FORM

There are two other forms of tableau representations which are in vogue (to represent the linear programming problem matrix) for hand computation or simple computer methods for the solution of linear programs.

Consider the problem in (2.3) restated as

$$\begin{aligned} &\text{Maximize } z = 0 + c_1x_1 + c_2x_2 + \ldots + c_nx_n \\ &\text{subject to } a_{11}x_1 + a_{12}x_2 + \ldots + a_{1n}x_n + x_{n+1} = b_1 \\ &\qquad a_{m1}x_1 + a_{m2}x_2 + \ldots + a_{mn}x_n + \ldots + x_{n+m} = b_m \end{aligned} \qquad (2.31)$$

and

$$x_1, x_2, \ldots, x_{n+m} \geqslant 0.$$

This may now be represented in the Extended Tableau Format including the unit columns, for $x_{n+1}, x_{n+2}, \ldots, x_{n+m}$.

Basic variables	r.h.s.	x_1	x_2	x_n	z	x_{n+1}	x_{n+m}
z	0	$-c_1$	$-c_2$	$-c_n$	1	0	0
x_{n+1}	b_1	a_{11}	a_{12}	a_{1n}	0	1	0
x_{n+2}	b_2	a_{21}	a_{22}	a_{2n}	0	0	0
$\vdots$							
x_{n+m}	b_m	a_{m1}	a_{m2}	a_{mn}	0	0	1

EXTENDED TABLEAU

In this representation a typical row may be read as,

$$b_1 = a_{11}x_1 + a_{12}x_2 + \ldots + a_{1n}x_n + x_{n+1},$$

and the objective row may be read as

$$0 = -c_1x_1 - c_2x_2 - \ldots - c_nx_n + z.$$

Note that z now onwards may always be treated as a basic variable, the only difference being this variable may take positive or negative value and shall never leave the basis.

In the Tucker–Beale form of representation $m + n$, x variables and the z variable are expressed as row equations in terms of the n non basic variables: this is set out in the following tableau.

TUCKER–BEALE FORM

		$-x_1$	$-x_2$	$\ldots$	$-x_n$
z	0	$-c_1$	$-c_2$	$\ldots$	$-c_n$
x_{n+1}	b_1	a_{11}	a_{12}	$\ldots$	a_{1n}
x_{n+2}	b_2	a_{21}	a_{22}	$\ldots$	a_{2n}
$\vdots$	$\vdots$	$\vdots$	$\vdots$		$\vdots$
x_{n+m}	b_m	a_{m1}	a_{m2}	$\ldots$	a_{mn}
x_1	0	-1	0	$\ldots$	0
x_2	0	0	-1	$\ldots$	0
$\vdots$	$\vdots$	$\vdots$	$\vdots$		$\vdots$
x_n	0	0	0		-1

In the extended tableau there are always m unit columns and in the Tucker–Beale Form there are n-unit rows. When pivotal operations are carried out in the contracted tableau variables exchange their row and column positions, in the extended tableau they stay in their column home positions and in the Tucker–Beale Form in their row home positions.

EXERCISES

Deduce the following rules for pivotal transformation for the extended tableau and Tucker–Beale Form; $\bar{a}_{pq} \neq 0$ is the pivotal element in the pth row and qth column.

Extended tableau

$$\begin{aligned}
\bar{a}'_{pq} &= 1 \\
\bar{a}'_{iq} &= 0; \qquad i = 0, 1, 2, \ldots, m \quad \text{and} \quad i \neq p \\
\bar{a}'_{pj} &= \bar{a}_{pj}/\bar{a}_{pq}; \qquad j = 0, 1, \ldots, n + m \quad \text{and} \quad j \neq q \\
\bar{a}'_{ij} &= \bar{a}_{ij} - \bar{a}_{iq} . \bar{a}_{pj}/\bar{a}_{pq} \text{ for all } i, j \text{ except } i = p \text{ or } j = q.
\end{aligned} \tag{2.32}$$

Tucker–Beale form

$$\begin{aligned}
\bar{a}'_{pq} &= -1 \\
\bar{a}'_{iq} &= -\bar{a}_{iq}/\bar{a}_{pq}; \qquad i = 0, 1, \ldots, n + m \quad \text{and} \quad i \neq p \\
\bar{a}'_{pj} &= 0; j = 0, 1, \ldots, n \quad \text{and} \quad j \neq q \\
\bar{a}'_{ij} &= \bar{a}_{ij} - \bar{a}_{iq}\bar{a}_{pj}/\bar{a}_{pq} \text{ for all } i, j \text{ except } i = p \text{ or } j = q.
\end{aligned} \tag{2.33}$$

2.4 Solve the problem

Minimize $\qquad 35x_1 + 40x_2 + 30x_3 + 55x_4 + 45x_5$

subject to

$$x_1 + x_2 + x_3 + x_4 + x_5 = 1$$
$$10x_1 + 15x_2 + 20x_3 + 35x_4 + 40x_5 \geqslant 25,$$

and

$$x_1, x_2 \ldots x_5 \geqslant 0.$$

2.5 Solve the problem

Minimize $\qquad 3x_1 + 4x_2 + 5x_3$

subject to

$$x_1 + x_2 + x_3 = 1$$

$$5x_2 + 3x_3 \geqslant 2$$

$$x_2 + 2x_3 \geqslant 1$$

$$4x_2 + x_3 \leqslant 2$$

and

$$x_1, x_2, x_3 \geqslant 0$$

2.6 Set up the following problem in the contracted tableau form and apply the simplex algorithm.

Maximize

$$2x_1 + x_2$$

subject to

$$-x_1 + x_2 \leqslant 2$$

$$x_1 - 3x_2 \leqslant 3$$

$$x_1 - 2x_2 \leqslant 4$$

and

$$x_1, x_2 \geqslant 0.$$

What is the exit condition from the algorithm? Verify your result by a graphical representation of the problem and the solution steps.

CHAPTER 3

Some Ancillary Features of the Simplex Method

KEYWORDS: Phase 1, Phase 2, Infeasibility Form, Artificial Variable, Primal Degeneracy, Cycling.

3.1 METHOD OF OBTAINING A FEASIBLE SOLUTION TO A LINEAR PROGRAM

The simplex method works on the assumption, that one starts with a solution that is basic, and also feasible (i.e. solution values of the basic variables are non-negative hence satisfy the constraints). Moves can then be made via the neighbouring bases always maintaining feasibility and improving the objective function value until the optimum basic feasible solution is obtained.

Initially, if a basic feasible solution is not at hand a method has to be devised to obtain one: towards this purpose the simplex method itself may be re-applied as follows.

For an inequality of the form

$$a_{11}x_1 + a_{12}x_2 + \ldots + a_{1n}x_n \leqslant -b_1, \tag{3.1}$$

if a non-negative slack variable x_{n+1} is introduced, this results in the equality,

$$a_{11}x_1 + a_{12}x_2 + \ldots + a_{1n}x_n + x_{n+1} = -b_1,$$

or

$$x_{n+1} = -b_1 - (a_{11}x_1 + a_{12}x_2 + \ldots + a_{1n}x_n), \tag{3.2}$$

then a feasible starting point cannot be obtained by setting other variables to zero, as $x_{n+1} = -b_1$ whereas it should be nonnegative. Note that changing the sign in (3.1) and hence (3.2) does not improve matters either. However, if a second variable x_{t1} called the "Artificial Variable" is introduced for this constraint (it is postulated that in any feasible solution such a variable

B

must take the value zero) then (3.1) can be rewritten as

$$a_{11}x_1 + a_{12}x_2 + \ldots + a_{1n}x_n + x_{n+1} - x_{t1} = -b_1,$$

or

$$x_{t1} = b_1 + a_{11}x_1 + a_{12}x_2 + \ldots + a_{1n}x_n + x_{n+1}. \tag{3.3}$$

In this case $x_{t1} = b_1$ is a feasible starting value for an auxiliary problem which must be first solved to obtain a feasible starting point of the original problem of which (3.1) is a constraint.

For an equality of the form

$$a_{i1}x_1 + a_{i2}x_2 + \ldots + a_{in}x_n = b_i \tag{3.4}$$

an artificial variable x_{ti} is introduced such that

$$a_{i1}x_1 + x_{i2}x_2 + \ldots + a_{in}x_n + x_{ti} = b_i$$

again this variable must be reduced to zero.

A student trying to work out for the first time the rationale of introducing artificial variables, often wonders why a negative pivot $a_{1p} < 0$ in (3.2) and a positive pivot $a_{ij} > 0$ in (3.4) are not used to obtain the first feasible non-negative values for the variable x_p and x_j. This is not done because when the values of x_p and x_j are substituted in the other equations this might lead to the introduction of negative solution values as no simplex ratio rules have been observed.

Consider the problem:

$$\text{Maximize } z = c_1x_1 + c_2x_2 + \ldots + c_nx_n, \tag{3.5}$$

subject to the general system of constraints,

$$\begin{aligned}
a_{11}x_1 + a_{12}x_2 + \ldots + a_{1n}x_n &\geqslant b_1 \\
a_{21}x_1 + a_{22}x_2 + \ldots + a_{2n}x_n &= b_2 \\
\cdot \qquad \cdot \qquad \cdot & \\
a_{31}x_1 + a_{32}x_2 + \ldots + a_{3n}x_n &- \leqslant b_3 \\
\vdots \qquad \vdots \qquad \ldots \quad \vdots & \\
a_{m1}x_1 + a_{m2}x_2 + \ldots + a_{mn}x_n &\leqslant b_m,
\end{aligned} \tag{3.6}$$

and

$$x_1, x_2, \ldots, x_n \geqslant 0.$$

The slack variables $x_{n+1}, x_{n+3}, \ldots, x_{n+m}$ and the artificial variables x_{t1} and x_{t2} are introduced and the equivalent system of equations,

$$
\begin{aligned}
-(a_{11}x_1 + a_{12}x_2 + \ldots + a_{1n}x_n + x_{n+1}) + x_{t1} &= b_1 \\
a_{21}x_1 + a_{22}x_2 + \ldots + a_{2n}x_n \qquad + x_{t2} &= b_2 \\
\cdot \qquad \cdot \qquad \cdot \qquad & \\
a_{31}x_1 + a_{32}x_2 + \ldots + a_{3n}x_n \qquad + x_{n+3} &\leqslant b_3 \\
\cdot \qquad \cdot \qquad \cdot \qquad & \\
a_{m1}x_1 + a_{m2}x_2 + \ldots + a_{mn}x_n \qquad + x_{n+m} &= b_m
\end{aligned}
\tag{3.7}
$$

is obtained where $x_1, x_2, \ldots, x_n, x_{n+1}, x_{n+3}, \ldots, x_{n+m}$ must take non-negative values and x_{t1}, x_{t2} must take zero values.

If an "Infeasibility Form" w,

$$
w = x_{t1} + x_{t2}, \tag{3.8}
$$

is now minimized subject to the constraints (3.7) and w reduces to zero, a feasible solution to the original constraint set in (3.6) is obtained. If w cannot be reduced to zero, "No Feasible" solution exists to the constraint set in (3.6). The simplex steps by which the infeasibility form w is minimized are called the "Phase 1" steps. At the end of the Phase 1 steps (when $w = 0$) one switches over to the normal simplex method and this is known as the "Phase-2" of the simplex method; in Phase 2 the objective function z of the original problem is minimized.

3.2 PHASE 1 AND PHASE 2 OF THE SIMPLEX METHOD: AN EXAMPLE

Consider the problem,

$$
\begin{aligned}
\text{Maximize } z &= 0 + 3x_1 + 2x_2 + x_3 + 2x_4 \\
\text{subject to} \quad & 3x_1 + 4x_2 + 5x_3 + 4x_4 \leqslant 5 \\
& 2x_1 + 6x_2 + x_3 + 5x_4 \geqslant 6 \\
& x_1 + x_2 + 5x_3 + 2x_4 = 2
\end{aligned}
\tag{3.9}
$$

and

$$
x_1, x_2, x_3, x_4 \geqslant 0.
$$

Introduce slack variables x_5 and x_6, artificial variables x_{t1} and x_{t2} and write the equations (3.9) as

Maximize z,

$$
\begin{aligned}
-3x_1 - 2x_2 - x_3 - 2x_4 + z \qquad &= 0 \\
\text{subject to} \quad 3x_1 + 4x_2 + 5x_3 + 4x_4 + x_5 \qquad &= 5 \\
2x_1 + 6x_2 + x_3 + 5x_4 - x_6 + x_{t1} &= 6 \\
x_1 + x_2 + 5x_3 + 2x_4 \qquad + x_{t2} &= 2,
\end{aligned}
\tag{3.10}
$$

Tableau	Basic variables	r.h.s.	x_1	x_2	x_3	x_4	x_5	x_6	x_{t1}	x_{t2}	$-w$	z
Tableau 3.1 Phase 1	$-w$	−8	−3	−7	−6	−7		1			1	
	z	0	−3	−2	−1	−2						1
	x_5	5	3	4	5	4	1					
	x_{t1}	6	2	6	1	5		−1	1			
	x_{t2}	2	1	1	5	2				1		
Tableau 3.2 Phase 1	$-w$	−1	−2/3		−29/6	−7/6		−1/6	7/6		1	
	z	2	−7/3		−2/3	−1/3		−1/3	1/3			1
	x_5	1	5/3		13/3	2/3	1	2/3	−2/3			
	x_2	1	1/3	1	1/6	5/6		−1/6	−1/6			
	x_{t2}	1	2/3		29/6	7/6		1/6	−1/6	1		
Tableau 3.3 Phase 1	$-w$	0							1	1	1	
	z	62/29	−65/29			−5/29		−9/29	9/29	4/29		1
	x_5	3/29	31/29			−11/29	1	15/29	−15/29	−26/29		
	x_2	28/29	9/29	1		23/29		−5/29	5/29	−1/29		
	x_3	6/29	4/29		1	7/29		1/29	−1/29	6/29		
Tableau 3.4 Phase 2	z	73/31				−30/31	65/31	24/31				1
	x_1	3/31	1			−11/31	29/31	15/31				
	x_2	29/31		1		28/31	−9/31	−10/31				
	x_3	6/31			1	9/31	−4/31	−1/31				
Tableau 3.5 Optimum	z	3			30/9		5/3	6/9				1
	x_1	1/3	1		11/9		7/9	4/9				
	x_2	1/3		1	−28/9		1/9	2/9				
	x_4	2/3			31/9	1	−4/9	−1/9				

where $x_1, x_2, \ldots, x_6$ must be non-negative and x_{t1} and x_{t2} must be zero. In order to drive x_{t1} and x_{t2} to zero the Phase 1 infeasibility form is obtained

$$\begin{aligned} w &= x_{t1} + x_{t2} \\ &= 6 - 2x_1 - 6x_2 - x_3 - 5x_4 + x_6 \\ &\quad + 2 - x_1 - x_2 - 5x_3 - 2x_4 \\ &= 8 + 3(-x_1) + 7(-x_2) + 6(-x_3) + 7(-x_4) - (-x_6) \end{aligned} \tag{3.11}$$

If one followed an early suggestion of Charnes [11] the expression for a composite objective function $z - M.\, w$ would be maximized where M is a very large number compared to the coefficients of the original objective function. However, a more accepted method is to first reduce the infeasibility form w to zero and then optimize the original objective function; inaccuracies due to heavily weighting some of the coefficients stored as floating point numbers are not introduced in this method.

The problem is set up in Tableau 3.1 which contains both the objective function row and the infeasibility form w. The simplex method is applied and a basic feasible solution with $w = 0$ is obtained in Tableau 3.3. From the following Tableau 3.4 the w-row and the artificial columns are dropped i.e. these are not updated during the tableau transformations. Pivot selections and the subsequent transformations are carried out in Tableaux 3.3 and 3.4 by the primal simplex rules for Phase 2; Tableau 3.5 contains the optimum solution.

3.3 SOME PRACTICAL CONSIDERATIONS FOR THE PHASE 1 COMPUTATION

If the coefficients of the infeasibility form and those in the columns of the artificial and slack variables are studied some interesting properties emerge. It is obvious from the way (3.11) the infeasibility form w, is obtained that the coefficients of this expression are the sum of the coefficients of the rows containing the artificial variables which are summed up. For instance the coefficient -3 of x_1 is the sum of two coefficients -2, -1 of the two rows containing x_{t1}, x_{t2} respectively. Consider now the second equation of (3.10) which contains two supplementary variables x_6 and x_{t1} with coefficients of the same magnitude (unity) but of opposite sign. As a consequence in a tableau in which one of these variables is basic the other has a negative unit vector, and in a tableau where both are non-basic these have column coefficients of the same magnitude but of opposite sign, except for the row in which $-w$ appears; observe columns of x_6 and x_{t1} Tableau 3.2, 3.3.

It is therefore not quite necessary either (i) to introduce the infeasibility form explicitly, or (ii) to introduce artificial variables for constraints which have their inequalities the wrong way round (3.2; 3.3); these information can be extracted from the rest of the tableau. This, however, implies that during Phase 1 non-negative slacks may be allowed to appear initially at an infeasible (negative) level. Let N_g denote the set of indices of those rows for which β_i the solution values are negative i.e. $N_g = \{i | \beta_i < 0\}$. Further let A_t denote the set of indices of those rows which specify equality yet to be satisfied i.e. artificial variables are contained in the basis at a positive level. Then the sum of the infeasibilities in this tableau may be computed by the relationship

$$S = \sum_{i \in N_g} \beta_i - \sum_{i \in A_t} \beta_i. \tag{3.12}$$

Since the infeasibility form is the linear sum and difference of the negative slack and positive artificial rows respectively the reduced cost coefficients or the $\bar{w}_j$'s of this form at any intermediate tableau may be given by the expression:

$$\bar{w}_j = \sum_{i=1}^{m} e_i \bar{a}_{ij} \tag{3.13}$$

where

$e_i = -1$ if x_{ti} (in the ith row) is artificial and $\beta_i > 0$ (A_t rows),

$e_i = 1$ if $\beta_i < 0$ and the variable in this row is slack, structural or artificial (N_g-rows),

$e_i = 0$ otherwise.

During the row choice of the simplex algorithm a negative pivot against a negative value of β_i is admitted in the ratio test. Under these circumstances the simplex pivot steps carried out in Phase 1 are the same as those which would otherwise be carried out using an explicit infeasibility row and a separate artificial variable for each individual infeasibility.

3.4 REDUNDANCY AND INCONSISTENCY IN THE CONSTRAINT EQUATIONS: DEGENERATE SOLUTIONS TO CONSTRAINT SETS

In this section some properties of the coefficient matrix, augmented by the r.h.s. vector are considered. It is assumed that the reader has some understanding of the rank of a matrix and the linear dependence of the rows and columns of a matrix. In case of difficulty the reader may refer to any standard text book on linear algebra or the chapter on mathematical background in [26].

Consider the Phase 1 of the simplex method in the not compact form, i.e. with an explicit infeasibility row and artificial variables one for every infeasible inequality and equality constraint. The Phase 1 may terminate in one of three possible ways.

(i) All the artificial variables have been successfully removed from the basis.
(ii) Artificial variables are contained in the basis at zero level.
(iii) Artificial variables are contained in the basis at a positive level.

In the first case it is clear that a basic feasible solution has been obtained; the derived equations and hence the original constraints are consistent and not redundant. The example of Section 3.2 illustrates this case.

In the second case also it is obvious that there exists a basic feasible solution to the constraint set. However, there may be two further outcomes of this case. Firstly if for a row i in which an artificial variable appears at a zero level the coefficients $\bar{a}_{ij}$ for all j excluding the artifical columns assume the value $\bar{a}_{ij} = 0$ then clearly this row is linearly dependent on all other rows of the transformed matrix. Indeed this represents a redundant constraint. On the other hand if for some j (where this is not one of the columns for artificial variables) $\bar{a}_{ij} \neq 0$ then clearly it is possible to pivot on this element and the variable x_j can enter the basis at zero level, $\bar{\beta}'_i = 0/\bar{a}_{ij}$, and the artificial variable can be removed.

Consider the transportation problem,

$$\text{Minimize } 20x_{11} + 30x_{12} + 25x_{13} + 25x_{21} + 35x_{22} + 40x_{23} + 15x_{31} + 25x_{32} + 35x_{33} \tag{3.14}$$

subject to

$$\begin{aligned}
x_{11} + x_{12} + x_{13} &= 25 \\
x_{21} + x_{22} + x_{23} &= 25 \\
x_{31} + x_{32} + x_{33} &= 35 \\
x_{11} \qquad x_{21} \qquad x_{31} &= 15 \\
x_{12} \qquad + x_{22} \qquad x_{32} &= 30 \\
x_{13} \qquad x_{23} \qquad x_{33} &= 40
\end{aligned}$$

and

$$x_{11}, x_{12} \ldots x_{33} \geqslant 0.$$

If the first three and the last three equations are added separately one gets the same relationship, i.e.

$$\sum_{i=1}^{3} \sum_{j=1}^{3} x_{ij} = 85.$$

It may be shown [26] that if any one row of the problem is dropped the feasible region defined by the constraint set still remains the same. The problem is set up in Tableau 3.6 and the first basic feasible solution is obtained in Tableau 3.11 .Observe that x_{t6} has no entry in the columns of any other variables in the problem. This redundant row continues to have artificial at zero level up to the optimal Tableau 3.15. This illustrates the case of the redundant constraints.

For an illustration of a degenerate solution to a linear program consider the problem

$$\begin{aligned} \text{Maximize } f(x) &= 5x_1 + 4x_2, \\ \text{subject to} \qquad & 2x_1 + 4x_2 \leqslant 1000 \\ & 2x_1 + x_2 \leqslant 400 \\ & x_1 \leqslant 200 \end{aligned} \tag{3.15}$$

and

$$x_1, x_2 \geqslant 0.$$

The problem is set up in a contracted Tableau 3.16. The tableau is primal feasible dual infeasible and the Phase 2 Primal algorithm is applied. Note that there is a tie for the variable to go out of the basis. The two sequences of tableaux to the optimal solution 3.16 → 3.17 → 3.18 or 3.16 → 3.19 → 3.20 → 3.18 are illustrated; these result from two optional pivots. In the second case there is an extra pivot step. In the first case there is one degenerate tableau in the second case there are two such degenerate intermediate tableaux. However, the real curse of degeneracy is cycling: some degenerate problems may be such that during simplex iterations bases may be repeated without any improvement of the objective function. Contrary to the statements made in earlier text books in this field, degenerate problems and cycling are common phenomena in the large scale models of present day. Computer codes must provide sophisticated algorithmic facilities for breaking ties under these conditions. The example shown here is a case of primal degeneracy. When the reduced cost coefficients of some of the non basic variables vanish this is known as dual degeneracy.

Returning to the last of the three cases mentioned earlier this happens when there can be "no feasible solution" to the constraint equations, i.e., the sum of infeasibilities cannot be reduced to zero. Let a procedure very similar to the one described for zero artificial be started i.e. for any row containing an artificial variable a non zero pivot $\bar{a}_{1j} \neq 0$ be sought and let the artificial be pivoted out if one such element is found. If all the artificials are pivoted out and in the basis only structural and slack variables are left at negative level

	Basic variables	r.h.s.	x_{11}	x_{12}	x_{13}	x_{21}	x_{22}	x_{23}	x_{31}	x_{32}	x_{33}	x_{t1}	x_{t2}	x_{t3}	x_{t4}	x_{t5}	x_{t6}
Tableau 3.6	$-w$	-170	-2	-2	-2	-2	-2	-2	-2	-2	-2						
	$-z$	0	20	30	25	25	35	40	15	25	35						
	x_{t1}	25	1	1	1							1					
	x_{t2}	25				1	1	1					1				
	x_{t3}	35							1	1	1			1			
	x_{t4}	15	(1)			1			1						1		
	x_{t5}	30		1			1			1						1	
	x_{t6}	40			1			1			1						1
Tableau 3.7	$-w$	-140		-2	-2		-2	-2		-2	-2				2		
	z	-300		30	25	5	35	40	-5	25	35				-20		
	x_{t1}	10		(1)	1	-1			-1			1			-1		
	x_{t2}	25				1	1	1					1				
	x_{t3}	35							1	1	1			1			
	x_{11}	15	1			1			1						1		
	x_{t5}	30		1			1			1						1	
	x_{t6}	40			1			1			1						1
Tableau 3.8	$-w$	-120				-2	-2	-2	-2	-2	-2	2					
	$-z$	-600			-5	35	35	40	25	25	35	-30			10		
	x_{12}	10		1	1	-1			-1			1			-1		
	x_{t2}	25				1	1	1					1				
	x_{t3}	35							1	1	1			1			
	x_{11}	15	1			1			1						1		
	x_{t5}	20			-1	1	(1)		1	1		-1			1	1	
	x_{t6}	40			1			1			1						1
Tableau 3.9	$-w$	-80			-2			-2			-2				2	2	
	$-z$	-1300			30			40	-10	-10	35	5			-25	-35	
	x_{12}	10		1	1	-1			-1			1			-1		
	x_{t2}	5			1			(1)	-1	-1		1	1		-1		
	x_{t3}	35							1	1	1			1			
	x_{11}	15	1			1			1						1		
	x_{22}	20			-1	1	1		1	1					1	1	
	x_{t6}	40			1			1			1						1

	Basic variables	r.h.s.	x_{11}	x_{12}	x_{13}	x_{21}	x_{22}	x_{23}	x_{31}	x_{32}	x_{33}	x_{t1}	x_{t2}	x_{t3}	x_{t4}	x_{t5}	x_{t6}
Tableau 3.10	$-w$	-70							-2	-2	-2	2	2			2	
	$-z$	-1500			-10				30	30	35	-35	-40		15	-35	
	x_{12}	10		1	1	-1			-1			1			-1		
	x_{23}	5			1			1	-1	-1		1	1		-1		
	x_{t3}	35							1	1	(1)			1			
	x_{11}	15	1			1			1						1		
	x_{22}	20			-1	1	1		1	1		-1			1	1	
	x_{t6}	35							1	1	1	-1	-1		1		1
Tableau 3.11	$-w$	0										2	2	2		2	
	$-z$	-2725			-10				-5	-5		-35	-40	-35	15	-35	
	x_{12}	10		1	1	-1			-1			1			-1		
	x_{23}	5			(1)			1	-1	-1		1	1		-1		
	x_{33}	35							1	1	1			1			
	x_{11}	15	1			1			1						1		
	x_{22}	20			-1	1	1		1	1		-1			1	1	
	x_{t6}	0										-1	-1	-1	1		1
Tableau 3.12	$-z$	-2675						10	-15	-15							
	x_{12}	5		1		-1		-1		1							
	x_{13}	5			1			1	-1	-1							
	x_{33}	35							1	1	1						
	x_{11}	15	1			1			(1)								
	x_{22}	25				1	1	1									
	x_{t6}	0															
Tableau 3.13	$-z$	-2450	15			15		10		-15							
	x_{12}	5		1		-1		-1		(1)							
	x_{13}	20	1		1			1		-1							
	x_{33}	20	-1							1	1						
	x_{31}	15	1			1			1								
	x_{22}	25				1	1	1									
	x_{t6}	0															

	Basic variables	r.h.s.	x_{11}	x_{12}	x_{13}	x_{21}	x_{22}	x_{23}	x_{31}	x_{32}	x_{33}
TABLEAU 3.14	$-z$	-2375	15	15				-5			
	x_{32}	5		1		-1		-1		1	
	x_{13}	25	1	1	1	-1					
	x_{33}	15	-1	-1		1		1			1
	x_{31}	15	1			1			1		
	x_{22}	25				1	1	1			
	x_{t6}	0									
TABLEAU 3.15	$-z$	-2300	10	10		5					5
	x_{32}	20	-1							1	1
	x_{13}	25	1	1	1	-1					
	x_{23}	15	-1	-1		1		1			1
	x_{31}	15	1			1			1		
	x_{22}	10					1				-1
	x_{16}	0									

	1	$-x_1$	$-x_2$
$f(x)$	0	-5	-4
x_3	1000	2	4
x_4	400	2	1
x_5	200	1	0

TABLEAU 3.16

	1	$-x_4$	$-x_2$
$f(x)$	1000	5/2	$-3/2$
x_3	600	-1	3
x_1	200	$\frac{1}{2}$	$\frac{1}{2}$
x_5	0	$-\frac{1}{2}$	$-\frac{1}{2}$

TABLEAU 3.17

	1	$-x_4$	$-x_3$
$f(x)$	1300	2	$\frac{1}{2}$
x_2	200	$-1/3$	1/3
x_1	100	2/3	$-1/6$
x_5	100	$-2/3$	1/6

TABLEAU 3.18

	1	$-x_5$	$-x_2$
$f(x)$	1000	5	-4
x_3	600	-2	4
x_4	0	-2	1
x_1	200	1	0

TABLEAU 3.19

	1	$-x_5$	$-x_4$
$f(x)$	1000	-3	4
x_3	600	6	-4
x_2	0	-2	1
x_1	200	1	0

TABLEAU 3.20

x_3 and x_4 columns are interchanged.

x_5 and x_1 row positions are also interchanged.

then there exist basic solutions to the original constraints but not a feasible one, however, if the above procedure fails because $\bar{a}_{ij} = 0$ for all j corresponding to a row with artificial variables at a positive level clearly the original set of equations is inconsistent. No feasible solutions are encountered in real life situations when there is some error in model formulation. A sensible LP system when it encounters this, outputs the tableau elements for the infeasible rows. This is often an invaluable aid for detecting such errors.

EXERCISES

3.1 Solve the problem in section 3.2 using only one artificial variable and without using any explicit infeasibility form. How should the simplex pivot choice rule be altered to apply this method?

3.2 Comment on the nature of difficulty experienced in solving the problems in Exercise 2.4 and 2.5. Solve these two problems applying the Phase 1 method.

3.3 What are the possible exits from the simplex algorithm? Describe the algorithm using step by step statements and discuss the conclusions which may be made corresponding to these different exit conditions.

3.4 Obtain a starting feasible solution to the problem

Minimize
$$x_2 - 3x_3 + x_4 + 2x_5$$

subject to
$$\begin{aligned} x_1 + 3x_2 - x_3 + 2x_5 &= 7 \\ -2x_2 + 4x_3 + x_4 &= 12 \\ -4x_2 + 3x_3 + 8x_5 + x_6 &= 10 \end{aligned}$$

and
$$x_1, x_2, \ldots, x_6 \geqslant 0.$$

Can you obtain a starting feasible solution without applying the Phase 1 of the simplex algorithm?

3.5 Obtain feasible solutions to the following three sets of constraints

(i)
$$\begin{aligned} 5x_2 + 3x_3 &\geqslant 2, \\ x_2 + 2x_3 &\geqslant 1, \\ 5x_2 + x_3 &\leqslant 2, \end{aligned}$$
and $x_1, x_2, x_3 \geqslant 0.$

(ii)
$$\begin{aligned} 2x_1 + 4x_2 + 6x_3 &= 5, \\ x_1 + 4x_2 + 4x_3 &= 3, \\ x_1 + 2x_2 + 3x_3 &= 3, \end{aligned}$$
and $x_1, x_2, x_3 \geqslant 0.$

(iii)
$$\begin{aligned} 2x_1 + x_2 &\geqslant 3, \\ x_1 + x_2 &\leqslant 1, \end{aligned}$$
and $x_1, x_2 \geqslant 0.$

Two of these problems are not solvable. Comment on the nature of the difficulty encountered in each of these two cases.

CHAPTER 4

The Revised Simplex Method

KEYWORDS: Basis Matrix, Pricing Vector, Column Transformation, Product Form of Inverse, Eta Vector, Forward Transformation, Backward Transformation.

4.1 WHY REVISED SIMPLEX?

The linear programming matrices in real life are thin and sparse, i.e., there are many more columns than rows, and most of the coefficients of the matrix are likely to be zeros. A typical problem with say 400 rows may be 1500 columns long and have 3000 elements, i.e. be only 0·5% dense.

If the information in the initial tableau and intermediate steps are closely scrutinised it is observed that an amount of redundant information is generated in each step; this is also true for the optimal (target) tableau. Indeed, the most sought after information in the optimal tableau are the reduced cost coefficients and the solution values, contained in it. With this in view and to reduce the computational burden and also to make more efficient use of the fast storage of a computer, the Revised Simplex Method was devised by W. Orchard Hays and G. B. Dantzig [15]. This method does not require that all the elements of the intermediate tableaux be explicitly generated: the elements of the intermediate tableaux are generated as and when required. This can only be achieved if one knows (a representation of) the inverse of the current basis matrix for each tableau; the method therefore also goes by the name of "Inverse Basis" method. Alex Orden [15] proposed a representation of the Inverse as a product of a sequence of transformation matrices. When this representation is used the method goes by the name of "Product Form Algorithm". Further progress in the compact representation of the inverse of the basis matrix has since been made. In Section 4.4 such representations and in Section 5.4 the computational implications of the Revised Simplex Method are discussed.

4.2 THEORY OF THE METHOD AND THE ALGORITHMIC STEPS

Consider the problem
Minimize

$$c_1x_1 + c_2x_2 + \ldots + c_Nx_N$$

subject to

$$\begin{aligned} a_{11}x_1 + a_{12}x_2 + \ldots + a_{1N}x_N &= b_1 \\ a_{21}x_1 + a_{22}x_2 + \ldots + a_{2N}X_N &= b_2 \\ \vdots \quad\quad \vdots \quad\quad\quad\quad \vdots \quad & \quad \vdots \\ a_{m1}x_1 + a_{m2}x_2 + \ldots + a_{mN}X_N &= b_m. \end{aligned} \tag{4.1}$$

In order to be consistent with the dimensions of the problem stated in canonical form (Chapter 2) let it be further assumed that $N = m + n$. The problem may be transformed by carrying out m steps of Gaussian elimination with back substitution; at the end of these steps the reduced matrix should contain an $m \times m$ unit matrix and the corresponding dependent variables are expressed in terms of n remaining variables whose column entries make up the rest ($m \times n$) of the matrix, the r.h.s. values are also simultaneously transformed. At the end of these steps the system of equations reduces to a canonical form.

In order to gain some notional advantage and insight it is convenient to include the objective function with the rest of the equations. Hence a new variable x_0 is introduced and the objective form is rewritten as

$$x_0 + \sum_{j=1}^{N} c_jx_j = b_0.$$

The system of equations in (4.1) can now be expressed in terms of an $(m + 1) \times (N + 1)$ matrix

$$Ax = b. \tag{4.2}$$

Let the matrix A be partitioned such that $A = (A' \vdots B)$; (4.2) can be then expressed as

$$(A' \vdots B)\begin{pmatrix} x' \\ \cdots \\ x'' \end{pmatrix} = b, \tag{4.3}$$

where $x' = (x_1, x_2, \ldots, x_n)^T$, and $x'' = (x_0, x_{n+1}, \ldots, x_{n+m})^T$.

Writing out in full this takes the form

$$
\begin{array}{l}
c_1x_1 + c_2x_2 + \ldots + c_nx_n + \vdots\ x_0 + c_{n+1}x_{n+1} + \ldots + c_Nx_N = b_0 \\
a_{11}x_1 + a_{12}x_2 + \ldots + a_{1n}x_n + \vdots\ 0 + a_{1n+1}x_{n+1} + \ldots + a_{1N}x_N = b_1 \\
a_{21}x_1 + a_{22}x_2 + \ldots + a_{2n}x_n + \vdots\ 0 + a_{2n+1}x_{n+1} + \ldots + a_{2N}x_N = b_2 \\
\vdots \qquad\qquad \vdots \qquad A' \qquad\qquad \vdots \qquad\qquad B \qquad\qquad \vdots \qquad \vdots \\
a_{m1}x_1 + a_{m2}x_2 + \ldots + a_{mn}x_n + \vdots\ 0 + a_{mn+1}x_{n+1} + \ldots + a_{mN}x_N = b_m.
\end{array} \tag{4.4}
$$

Note that B is an $(m + 1) \times (m + 1)$ square submatrix and A' represents the rest of the partitioned columns of A. If B is not singular, i.e. determinant of B is not zero and B^{-1} exists, then the equation system can be solved for only

$$Bx'' = b, \tag{4.5}$$

and B is said to constitute a basis matrix of the equality system (4.2), (4.3), (4.4). Premultiplying (4.3) and (4.4) by B^{-1}, it follows that

$$(B^{-1}A' \ \vdots\ B^{-1}B)\, x = B^{-1}b = \beta,$$

or

$$(B^{-1}A' \ \vdots\ I)\, x = \beta. \tag{4.6}$$

If $\bar{a}_{ij}$ denotes a transformed element of the matrix then (4.6) may be expressed as

$$
\begin{array}{l}
\bar{a}_{01}x_1 + \bar{a}_{02}x_2 + \ldots + \bar{a}_{0n}x_n + x_0 \qquad\qquad\qquad = \beta_0 \\
\bar{a}_{11}x_1 + \bar{a}_{12}x_2 + \ldots + \bar{a}_{1n}x_n + \qquad + x_{n+1} \qquad\qquad = \beta_1 \\
\bar{a}_{21}x_1 + \bar{a}_{22}x_2 + \ldots + \bar{a}_{2n}x_n + \qquad\qquad + x_{n+2} \qquad = \beta_2 \\
\vdots \qquad\qquad \vdots \qquad\qquad \vdots \\
\bar{a}_{m1}x_1 + \bar{a}_{m2}x_2 + \ldots + \bar{a}_{mn}x_n + \qquad\qquad\qquad + x_{n+m} = \beta_m.
\end{array} \tag{4.7}
$$

This illustrates the transformation to the canonical form stated earlier in the section; note that if in (4.3) $B = I = B^{-1}$ such a reduction is not necessary. Further observe that (4.6) and (4.7) fully represent the extended tableau.

Some pertinent properties of the matrix relations should now be observed.

(1) If the inverse matrix B^{-1} is at hand, a matrix product $B^{-1}a_j$ where a_j is a column of the original matrix produces the updated column $\bar{a}_j = B^{-1}a_j$, $\bar{a}_j$ being the transformed column (of index j) of the transformed matrix/ tableau.

(2) For the purpose of illustration consider a particular basis matrix B made up of the columns of the coefficients corresponding to the variable x_0, and

the last m variables $x_{n+1}, x_{n+2}, \ldots, x_{n+m}$. The matrix B and its inverse are set out below

$$B = \begin{bmatrix} 1 & c_{n+1} & \ldots c_{n+m} \\ 0 & a_{1n+1} & \ldots a_{1n+m} \\ \vdots & \vdots & \ldots \vdots \\ 0 & a_{mn+1} & \ldots a_{mn+m} \end{bmatrix};$$

$$B^{-1} = \begin{bmatrix} 1 & \pi_1 & \pi_2 \, \pi_m \\ 0 & \bar{a}_{1n+1} & \bar{a}_{1n+m} \\ \vdots & \vdots & \vdots \\ 0 & \bar{a}_{mn+1} & \bar{a}_{mn+m} \end{bmatrix} = \begin{bmatrix} \pi_{00}\pi_{01} & \pi_{0m} \\ \pi_{10}\pi_{11} & \pi_{1m} \\ \vdots \quad \vdots & \vdots \\ \pi_{m0}\pi_{m1} & \pi_{mm} \end{bmatrix} = \begin{bmatrix} \pi_{0\cdot} \\ \pi_{1\cdot} \\ \vdots \\ \pi_{m\cdot} \end{bmatrix} \tag{4.8}$$

Note that a tableau corresponding to the B^{-1} of (4.8) contains the elements of the inverse matrix; these elements are denoted by $\bar{a}_{in+i}$ in the tableau. However, to facilitate development of some mathematical relationships presented in this section the elements of the B^{-1} matrix are denoted by π_{ij} and the rows of this matrix are denoted by $\pi_{i.}$, $i = 0, 1, \ldots, m$. This B^{-1} matrix must contain the unit column corresponding to the variable x_0 and always occupying the basis. The first component of the transformed column vector $\bar{a}_j$ may therefore be expressed as

$$\begin{aligned} \bar{c}_j = \bar{a}_{0j} &= (\text{first row of } B^{-1}).(\text{column } a_j) \\ &= (1, \pi_{01}, \pi_{02}, \ldots, \pi_{0m}).(c_j, a_{1j}, a_{2j}, \ldots, a_{mj})^T \\ &= c_j + \sum_{i=1}^{m} a_{ij}, \pi_{0i} = c_j + z_j \end{aligned} \tag{4.9}$$

This is a linear expression for the reduced cost coefficient, made up of c_j, z_j, z_j is often called the imputed cost i.e., the contribution of the coefficients a_{ij} multiplied by the shadow price π_{0i} and summed over all i. In a maximizing problem c_j is taken with a negative sign and $\bar{c}_j = z_j - c_j$, then becomes the expression for the reduced cost coefficient.

In order to follow the inverse matrix (revised simplex) algorithm both in Phase 1 and in Phase 2 the following relationships need to be understood.

(a) The first row vector of B^{-1} may be obtained by the vector matrix product

$$e_0 B^{-1} = \pi_{0.}, \tag{4.10}$$

where $e_0 = (1, 0, 0, \ldots 0)$ is a unit vector with unity in the 0th position, and $\pi_{0.} = (\pi_{00}, \pi_{01}, \ldots \pi_{0m})$ is the vector of the shadow prices. Then the reduced cost coefficient $\bar{c}_j$ may be expressed as

$$\bar{c}_j = \pi_{0.}a_j \quad . \tag{4.11}$$

(b) Now consider the problem of obtaining the sth component of the updated column $\bar{a}_j$, i.e. the element $\bar{a}_{sj}$. This is of course given by the product

$$\bar{a}_{sj} = \pi_{s.}a_j, \tag{4.12}$$

where $\pi_{s.}$ is the sth row of B^{-1} (see (4.8)). $\pi_{s.}$ therefore is obtained by the relationship

$$e_s B^{-1} = \pi_{s.}, \tag{4.13}$$

where $e_s = (0, 0, \ldots, 1, 0)$ has unity in the sth position and zero elsewhere.

(c) These ideas can be taken one step further to obtain the difference of say two elements $\bar{a}_{sj}, \bar{a}_{pj}$ of the updated column $\bar{a}_j$, It follows from (4.12) that

$$(\pi_{s.} - \pi_{p.})\, a_j = \bar{a}_{sj} - \bar{a}_{pj} \quad . \tag{4.14}$$

Hence

$$(e_s - e_p)\, B^{-1}.a_j = \bar{a}_{sj} - \bar{a}_{pj} \quad . \tag{4.15}$$

The expression $(e_s - e_p)$ can be combined to obtain one vector $e_{s-p} = (0, 0, \ldots, 1, \ldots, -1, 0, 0)$ i.e. with $+1$ in the sth and -1 in the pth position. The vectors e_0, e_s, e_p, e_{s-p} etc. which are used to obtain a row or sum or difference of rows of B^{-1} are called "form vectors". The mechanisms illustrated in (4.10), (4.11), and (4.14), (4.15) are used to combine the Phase 1 and Phase 2 of the Revised Simplex Method within a unified sequence of steps.

(3) It is not necessary to calculate from scratch every new inverse Basis B^{-1}. Let $(B^k)^{-1}$ denote the inverse basis at the kth iteration and let $\bar{a}_q$ be the updated column corresponding to the variable x_q which is being pivoted into the basis at the pth row position. Then the new inverse basis is given by the relationship.

$$(B^{k+1})^{-1} = T_k.(B^k)^{-1}$$

or

$$\begin{bmatrix} \pi_{00}^{k+1} & & \pi_{0m}^{k+1} \\ & & \\ & & \\ \pi_{m0}^{k+1} & & \pi_{mm}^{k+1} \end{bmatrix} = \begin{bmatrix} 1 & & -\bar{a}_{0q}/\bar{a}_{pq} & \\ & 1 & -\bar{a}_{1q}/\bar{a}_{pq} & \\ & & 1/\bar{a}_{pq} & \\ & & -\bar{a}_{mq}/\bar{a}_{pq} & 1 \end{bmatrix} . \begin{bmatrix} \pi_{00}^{k} & & \pi_{0m}^{k} \\ & & \\ & & \\ \pi_{m0}^{k} & & \pi_{mm}^{k} \end{bmatrix}$$

This operation is in fact identical with the pivotal transformation of the tableau method but carried out only on the inverse of the basis matrix.

Assuming that one works with an explicitly stored inverse matrix, the steps of the revised simplex method may be stated as:

Step 1. *Obtain π vector.*

In the first place obtain the π vector by which the reduced cost coefficient of a variable (column) will be computed. In Phase 2 it is the first row of the inverse matrix B^{-1}. In Phase 1, however, such a vector is the sum (weighted by $+1$ or -1) of the infeasibility rows of B^{-1}. π is therefore obtained by the vector premultiplication $\pi = eB^{-1}$, where in Phase 2,

$$e = (1, 0, 0, \ldots, 0),$$

and in Phase 1,

$$e = (0, 1, 0, -1, 0, \ldots),$$

i.e., the form vector is constructed by the rule given in Section 3, Chapter 3.

Step 2. *Pricing operation*

Price out the variables (columns of A) by computing the product $\pi a_j = \bar{a}_{0j}$, $j = 1, 2, \ldots, N$ and $j \notin$ basis. Choose a negative or the most negative $\bar{a}_{0q} = \bar{c}_q$; then q is the column to bring into the basis. If no such column exists go to Exit (In Phase 1 this means that "no feasible" solution exists; in Phase 2, the optimum solution has been attained).

Step 3. *Transform column*

Update the column q; $\bar{a}_q = B^{-1} a_q$ by the matrix by vector product indicated.

Step 4. *Choose a row* (*the corresponding variable to leave the basis*).

Perform a ratio test between the elements of β and $\bar{a}_q$ the updated r.h.s. and the incoming variable column and choose the pivotal row p:

$$\theta_p = \frac{\beta_p}{\bar{a}_{pq}} = \left\{\min_i \frac{\beta_i}{\bar{a}_{iq}} \geqslant 0;\, i = 1, 2, \ldots, m \text{ and } \bar{a}_{iq} \neq 0\right\}. \qquad (4.16)$$

Note that the ratio must always be non negative i.e. β_p must be of the same sign as the pivot element $\bar{a}_{pq}$: a proviso for the implicit phase 1 computation. In Pase 2 therefore ($\beta \geqslant 0$) the condition $\bar{a}_{iq} \neq 0$ implies that $\bar{a}_{iq} > 0$. If such a pivot does not exist in Phase 2 the problem is then declared 'Unbounded'; the control is transferred to Exit.

Step 5. Update the solution value and inverse matrix.

The solution values and the inverse matrix should be updated by the relationships

$$\begin{bmatrix} \beta_0^{k+1} \\ \beta_1^{k+1} \\ \\ \beta_m^{k+1} \end{bmatrix} = \begin{bmatrix} 1 & & -\bar{a}_{0q}/\bar{a}_{pq} & & \\ & 1 & -\bar{a}_{1q}/\bar{a}_{pq} & & \\ & & 1/\bar{a}_{pq} & 1 & \\ & & -\bar{a}_{mq}/\bar{a}_{pq} & & 1 \end{bmatrix} \cdot \begin{bmatrix} \beta_0^k \\ \beta_1^k \\ \\ \beta_m^k \end{bmatrix}, \tag{4.17}$$

and

$$\begin{bmatrix} \pi_{00}^{k+1} & \pi_{0m}^{k+1} \\ \\ \\ \pi_{m0}^{k+1} & \pi_{mm}^{k+1} \end{bmatrix} = \begin{bmatrix} 1 & & -\bar{a}_{0q}/\bar{a}_{pq} & & \\ & 1 & -\bar{a}_{1q}/\bar{a}_{pq} & & \\ & & \vdots & & \\ & & 1/\bar{a}_{pq} & & \\ & & \vdots & 1 & \\ & & -\bar{a}_{mq}/\bar{a}_{pq} & & 1 \end{bmatrix} \cdot \begin{bmatrix} \pi_{00}^{k} & \pi_{0m}^{k} \\ \\ \\ \pi_{m0}^{k} & \pi_{mm}^{k} \end{bmatrix} \tag{4.18}$$

Note that the transformations are the same as the pivotal transformations of Chapter 2 but are only expressed in matrix notation. After carrying out the transformations go back to Step 1.

EXIT: Control is transferred to this point under three conditions "no-feasible", "unbounded" and "optimal" solution.

4.3 AN EXAMPLE ILLUSTRATING THE STEPS OF THE REVISED SIMPLEX METHOD

Consider the problem illustrated in Section 2 of Chapter 3. The problem may be set up with only one artificial and two slack variables one of the latter taking negative starting value initially. But this practical phase 1 method is not considered in this section and this discussion is postponed until Chapter 5, Section 2. As in Chapter 3 two artificial variables are introduced and the product form steps of the illustrative example match one for one with the Tableau 3.1 through to Tableau 3.5.

The infeasibility form $w(x)$ is not explicitly created, therefore the basis will contain four vectors and is 4×4 in size.

In each of the five iterations the following information is noted: the form vector e, the pricing vector π^k, the basis matrix B^k and its inverse $(B^k)^{-1}$, the list of variables in the basis and the order in which these are pivoted and some ancillary information. Note that the variable z is renamed as variable x_0 in this section.

Iteration 1

Basis variables x_0, x_5, x_{t1}, x_{t2}

$$B^1 = \begin{bmatrix} 1 & & & \\ & 1 & & \\ & & 1 & \\ & & & 1 \end{bmatrix} = I\,(B^1)^{-1} = I = \begin{bmatrix} 1 & & & \\ & 1 & & \\ & & 1 & \\ & & & 1 \end{bmatrix} \quad (4.13)$$

Phase 1.

Form vector by the rule of (3.13) $e = (0, 0, -1\ \ -1)$; $\pi^1 = e(B^1)^{-1}$

$$= (0, 0, -1, -1) \begin{bmatrix} 1 & & & \\ & 1 & & \\ & & 1 & \\ & & & 1 \end{bmatrix} = (0, 0, -1, -1) \quad (4.14)$$

$$\text{Sum of infeasibility} = \pi^1 \times \beta^1 = (0, 0, -1, -1) \times \begin{bmatrix} 0 \\ 5 \\ 6 \\ 2 \end{bmatrix} = -8 \quad (4.15)$$

The reduced cost coefficients are

$$\overline{W}_{01} = \pi^1 \times \begin{bmatrix} -3 \\ 3 \\ 2 \\ 1 \end{bmatrix} = -3; \qquad \overline{W}_{02} = \pi^1 \times \begin{bmatrix} -2 \\ 4 \\ 6 \\ 1 \end{bmatrix} = -7;$$

$$\overline{W}_{03} = \pi^1 \times \begin{bmatrix} -1 \\ 5 \\ 1 \\ 5 \end{bmatrix} = -6; \qquad \overline{W}_{04} = \pi^1 \times \begin{bmatrix} -2 \\ 4 \\ 5 \\ 2 \end{bmatrix} = -7.$$

The reduced cost coefficients $\overline{W}_{06}$ and those of the basic variables are zero or $+1$. $\overline{W}_{02}$ is the first of the two most negative reduced cost coefficients so the variable x_2 is chosen to come into the basis. By the ratio test

$$\min\{\underset{i=2}{5/4}, \underset{i=3}{6/6}, \underset{i=4}{2/1}\},\ i = 3 \text{ and ratio} = 1,$$

it follows x_2 pivots into the third row and the transformation matrix becomes

$$T_1 = \begin{bmatrix} 1 & & 1/3 & \\ & 1 & -2/3 & \\ & & 1/6 & \\ & & -1/6 & 1 \end{bmatrix}. \tag{4.16}$$

β^2 and $(B^2)^{-1}$ are updated;

$$\beta^2 = T_1 . \beta^1 = \begin{bmatrix} 2 \\ 1 \\ 1 \\ 1 \end{bmatrix} \text{ and } (B^2)^{-1} = T_1(B^1)^{-1} = \begin{bmatrix} 1 & & 1/3 & \\ & 1 & -2/3 & \\ & & 1/6 & \\ & & -1/6 & 1 \end{bmatrix} \tag{4.17}$$

Iteration 2

Basic variables x_0, x_5, x_2, x_{t2}

$$B^2 = \begin{bmatrix} 1 & & -2 & \\ & 1 & 4 & \\ & & 6 & \\ & & 1 & 1 \end{bmatrix} \quad (B^2)^{-1} = \begin{bmatrix} 1 & & 1/3 & \\ & 1 & -2/3 & \\ & & 1/6 & \\ & & -1/6 & 1 \end{bmatrix} \tag{4.18}$$

Phase 1. Form vector $e = (0, 0, 0, -1); \pi^2 = e(B^2)^{-1} = (0, 0, 1/6, -1)$ (4.19)

$$\text{Sum of infeasibility} = \pi^2 \times \beta^1 = (0, 0, 1/6, -1) \times \begin{bmatrix} 0 \\ 5 \\ 6 \\ 2 \end{bmatrix} = -1 \tag{4.20}$$

The reduced cost coefficients are

$$\overline{W}_{01} = \pi^2 \times \begin{bmatrix} -3 \\ 3 \\ 2 \\ 1 \end{bmatrix} = -2/3; \qquad \overline{W}_{03} = \pi^2 \times \begin{bmatrix} -1 \\ 5 \\ 1 \\ 5 \end{bmatrix} = -29/6;$$

$$\overline{W}_{04} = \pi^2 \times \begin{bmatrix} -2 \\ 4 \\ 5 \\ 2 \end{bmatrix} = -7/6 \tag{4.21}$$

x_3 is now chosen to come into the basis. The updated entries in this column are,

$$(B^2)^{-1} \times a_3 = (B^2)^{-1} \times \begin{bmatrix} -1 \\ 5 \\ 1 \\ 5 \end{bmatrix} = \begin{bmatrix} -2/3 \\ 13/3 \\ 1/6 \\ 29/6 \end{bmatrix} = \bar{a}_3. \tag{4.22}$$

By the ratio test

$$\min_i \frac{\beta_i^2}{\bar{a}_{i3}} \Big\{ \underset{i=2}{1 \times 3/13}; \underset{i=3}{1 \times 6/1}, \underset{i=4}{1 \times 6/29} \Big\}\ i = 4 \text{ and ratio } 6/29;$$

it follows that x_3 pivots into row 4. The transformation matrix becomes.

$$T_2 = \begin{bmatrix} 1 & & & 4/29 \\ & 1 & & -26/29 \\ & & 1 & -1/29 \\ & & & 6/29 \end{bmatrix} \tag{4.23}$$

β^3 and $(B^3)^{-1}$ are updated

$$\beta^3 = T_2 \times \beta^2 = \begin{bmatrix} 62/29 \\ 3/29 \\ 28/29 \\ 6/29 \end{bmatrix}$$

$$\text{and} \quad (B^3)^{-1} = T_2 \times (B^2)^{-1} = \begin{bmatrix} 1 & & 9/29 & 4/29 \\ & 1 & -15/29 & -26/29 \\ & & 5/29 & -1/29 \\ & & -1/29 & 6/29 \end{bmatrix} \tag{4.24}$$

Iteration 3

Basic variables x_0, x_5, x_2, x_3

$$B^3 = \begin{bmatrix} 1 & & -2 & -1 \\ & 1 & 4 & 5 \\ & & 6 & 1 \\ & & 1 & 5 \end{bmatrix} \quad (B^3)^{-1} = \begin{bmatrix} 1 & & 9/29 & 4/29 \\ & & -15/29 & 26/29 \\ & & 5/29 & -1/29 \\ & & -1/29 & 6/29 \end{bmatrix}. \tag{4.25}$$

Phase 2.; all the artificial variables have been removed. Form vector for Phase 2:

$$e = (1, 0, 0, 0);\ \pi^3 = e \times (B^3)^{-1} = (1, 0, 9/29, 4/29) \tag{4.26}$$

The reduced cost coefficients are

$$\bar{c}_1 = \pi^3 \times \begin{bmatrix} -3 \\ 3 \\ 2 \\ 1 \end{bmatrix} = -65/29; \qquad \bar{c}_4 = \pi^3 \times \begin{bmatrix} -2 \\ 4 \\ 5 \\ 2 \end{bmatrix} = -5/29;$$

$$\bar{c}_6 = \pi^3 \times \begin{bmatrix} 0 \\ 0 \\ -1 \\ 0 \end{bmatrix} = -9/29.$$

Hence x_1 is chosen to come into the basis. The updated entries in this column are

$$(B^3)^{-1} a_1 = (B^3)^{-1} \times \begin{bmatrix} -3 \\ 3 \\ 2 \\ 1 \end{bmatrix} = \begin{bmatrix} -65/29 \\ 31/29 \\ 9/29 \\ 4/29 \end{bmatrix} = \bar{a}_1. \tag{4.27}$$

By ratio test

$$\min \frac{\beta_i^3}{a_{i1}} \{3/31, 28/9, 6/4\} \; i = 2, \text{ and ratio } 3/31$$

and it follows that x_1 pivots into row 2. The transformation matrix becomes.

$$T_3 = \begin{bmatrix} 1 & 65/31 & & \\ & 29/31 & & \\ & -9/31 & 1 & \\ & -4/31 & & 1 \end{bmatrix} \tag{4.28}$$

β^4 and $(B^4)^{-1}$ are updated.

$$\beta^4 = T_3 * \beta^3 = \begin{bmatrix} 73/31 \\ 3/31 \\ 29/31 \\ 6/31 \end{bmatrix};$$

$$(B^4)^{-1} = T_3 . (B^3)^{-1} = \begin{bmatrix} 1 & 65/31 & -24/31 & -54/31 \\ 0 & 29/31 & -15/31 & -26/31 \\ 0 & -9/31 & 10/31 & 7/31 \\ 0 & -4/31 & 1/31 & 10/31 \end{bmatrix} \tag{4.29}$$

Iteration 4

Basis variables x_0, x_1, x_2, x_3

$$B^4 = \begin{bmatrix} 1 & -3 & -2 & -1 \\ 0 & 3 & 4 & 5 \\ 0 & 2 & 6 & 1 \\ 0 & 1 & 1 & 5 \end{bmatrix}$$

$$(B^4)^{-1} = \begin{bmatrix} 1 & 65/31 & -24/31 & -54/31 \\ 0 & 29/31 & -15/31 & -26/31 \\ 0 & -9/31 & 10/31 & 7/31 \\ 0 & -4/31 & 1/31 & 19/31 \end{bmatrix} \tag{4.30}$$

Phase 2

Form vector $e = (1, 0, 0, 0)$; $\pi^4 = e.(B^4)^{-1}$

$$= (1, 65/31, -24/31, -54/31) \tag{4.31}$$

The reduced cost coefficients are

$$\bar{c}_4 = \pi^4 \times \begin{bmatrix} -2 \\ 4 \\ 5 \\ 2 \end{bmatrix} = -30/31;$$

$$\bar{c}_5 = \pi^4 \times \begin{bmatrix} 0 \\ 1 \\ 0 \\ 0 \end{bmatrix} = \frac{65}{51}; \qquad \bar{c}_6 = \pi^4 \times \begin{bmatrix} 0 \\ 0 \\ -1 \\ 0 \end{bmatrix} = 24/31.$$

Hence x_4 is chosen to come into the basis. The updated entries in this column are

$$(B^4)^{-1}.a_4 = (B^4)^{-1}. \begin{bmatrix} -2 \\ 4 \\ 5 \\ 2 \end{bmatrix} = \begin{bmatrix} -30/31 \\ -11/31 \\ 28/31 \\ 9/31 \end{bmatrix} = \bar{a}_4 \tag{4.32}$$

By the ratio test

$$\min_{\bar{a}_{i4} > 0} \frac{\beta_i^4}{\bar{a}_{i4}} \left\{ \frac{29}{28}, 6/9 \right\} i = 4 \text{ ratio } 6/9 \tag{4.33}$$

and it follows that x_4 pivots into row 4; the transformation matrix becomes

$$T_4 = \begin{bmatrix} 1 & & & 30/9 \\ & 1 & & 11/9 \\ & & 1 & -28/9 \\ & & & 31/9 \end{bmatrix}, \tag{4.34}$$

$$\beta^5 = T_4 \times \beta^4 = \begin{bmatrix} 3 \\ 1/3 \\ 1/3 \\ 2/3 \end{bmatrix}$$

$$(B^5)^{-1} = T_4(B^4)^{-1} = \begin{bmatrix} 1 & 5/3 & -6/9 & -2/3 \\ 0 & 7/9 & -4/9 & -4/9 \\ 0 & 1/9 & -2/9 & -7/9 \\ 0 & -4/9 & 1/9 & 10/9 \end{bmatrix} \tag{4.35}$$

Iteration 5

Basis variables x_0, x_1, x_2, x_4

$$B^5 = \begin{bmatrix} 1 & -3 & -2 & -2 \\ 0 & 2 & 4 & 4 \\ 0 & 2 & 6 & 5 \\ 0 & 1 & 1 & 2 \end{bmatrix} \quad (B^5)^{-1} = \begin{bmatrix} 1 & 5/3 & -6/9 & -2/3 \\ 0 & 7/9 & -4/9 & -4/9 \\ 0 & 1/9 & -2/9 & -7/9 \\ 0 & -4/9 & 1/9 & 10/9 \end{bmatrix} \tag{4.36}$$

Phase 2. Form vector

$$e = (1, 0, 0, 0,); \pi^5 = e.(B^5)^{-1} = (1, 5/3, -6/9, -2/3)$$

The reduced cost coefficient of the non basic variables are

$$\bar{c}_3 = \pi^5 \times \begin{bmatrix} -1 \\ 5 \\ 1 \\ 5 \end{bmatrix} = 10/3; \qquad \bar{c}_5 = \pi^5 \times \begin{bmatrix} 0 \\ 1 \\ 0 \\ 0 \end{bmatrix} = 5/3;$$

$$\bar{c}_6 = \pi^5 \times \begin{bmatrix} 0 \\ 0 \\ -1 \\ 0 \end{bmatrix} = 6/9$$

Since all the reduced cost coefficients are non negative this implies that the optimum solution is reached. The optimum solution value is given by β^5.

$$
\begin{aligned}
x_0 &= 3, \\
x_1 &= 1/3, \\
x_2 &= 1/3, \\
x_4 &= 2/3.
\end{aligned}
\tag{4.37}
$$

4.4 THE PRODUCT FORM OF THE INVERSE

The method of inverting a matrix by premultiplying it by a series of elementary transformation matrices is known and practised by numerical analysts over the years. This technique has since been adopted and sharpened by the algorithm designers for linear programming to extend the capability i.e., the size of problems which may be tackled on a given computer configuration.

Consider for the moment only the representation of the inverse of a matrix, and take the system of equations

$$Bx = \beta, \tag{4.38}$$

where B is a square $m \times m$ matrix. Assume further that it is possible to pivot down the diagonal of the matrix at each step of a series of m transformations. Such a process inverts the matrix and if the right hand side is updated solves the system of equations.

Consider now the matrix B^0

$$B^0 = (B \vdots I), \tag{4.39}$$

elements of this matrix may be considered as the elements of an extended tableau. Premultiplying this by the transformation matrix T_1

$$T_1(B^0) = T_1(B \vdots I) = B^1, \tag{4.40}$$

where

$$T_1 = \begin{bmatrix} \frac{1}{b_{11}} & 1 & & \\ \frac{-b_{21}}{b_{11}} & & 1 & \\ \frac{-b_{m1}}{b_{11}} & & & 1 \end{bmatrix} \quad \text{pivotal element is } b_{11} \tag{4.41}$$

This is indeed equivalent to carrying out a pivotal operation in the extended tableau with the pivot element $b_{11} \neq 0$. Now let $b^1_{22} \neq 0$ from B^1 be chosen as pivotal elements and one obtains the transformation matrix

$$T_2 = \begin{bmatrix} 1 & -\dfrac{b^1_{12}}{b'_{22}} & & & \\ & \dfrac{1}{b'_{22}} & 1 & & \\ & -\dfrac{b'_{m2}}{b'_{22}} & & 1 & \\ & & & & 1 \end{bmatrix} \tag{4.42}$$

and the relationship

$$T_2(B^1) = B^2,$$

or

$$T_2 T_1(B^0) = T_2 T_1(B \vdots I) = B^2. \tag{4.43}$$

The matrix relationships (4.40) and (4.43) may be written out in full as

$$\begin{bmatrix} \dfrac{1}{b_{11}} & & & \\ -\dfrac{b_{21}}{b_{11}} & 1 & & \\ & & 1 & \\ -\dfrac{b_{m1}}{b_{11}} & & & 1 \end{bmatrix} \times \begin{bmatrix} b_{11} \cdots b_{1m} & 1 & & \\ b_{21} \cdots b_{2m} & & 1 & \\ b_{m1} \cdots b_{mm} & & & 1 \end{bmatrix}$$

$$= \begin{bmatrix} 1 & b^1_{12} \cdots b^1_{1m} & \dfrac{1}{b_{11}} & 0 & & \\ 0 & b^1_{22} \cdots b^1_{2m} & -\dfrac{b_{21}}{b_{11}} & 1 & & \\ & & & & 1 & \\ 0 & b^1_{m2} \cdots b^1_{mm} & -\dfrac{b_{m1}}{b_{11}} & 0 & & 1 \end{bmatrix} \tag{4.44}$$

and

$$\begin{bmatrix} 1-\dfrac{b^1_{12}}{b^1_{22}} & & \\ \dfrac{1}{b^1_{22}} & 1 & \\ \dfrac{b^1_{m2}}{b^1_{22}} & & 1 \end{bmatrix} \times \begin{bmatrix} 1 & b^1_{12} & b^1_{1m} \cdots b^1_{1m+1} & 0 & 0 \\ & b^1_{22} & b^1_{2m} \cdots b^1_{2m+1} & 1 & \\ 0 & b^1_{m2} & b^1_{mm} \cdots b^1_{mm+1} & 0 & 1 \end{bmatrix}$$

$$= \begin{bmatrix} 1 & 0 & b^2_{13} \ldots b^2_{1m+1} & b^2_{1m+2} & 0 & & \\ 0 & 1 & b^2_{23} \ldots b^2_{2m+1} & b^2_{2m+2} & 0 & & \\ & & & & 1 & & \\ & & & & & 1 & \\ 0 & 0 & b^2_{m3} \ldots b^2_{mm+1} & b^2_{mm+2} & 0 & & 1 \end{bmatrix}. \qquad (4.45)$$

If this process is carried out through m steps then at the mth step of the transformation the matrix B^m will be

$$B^m = (I \vdots B^{-1}) = T_m T_{m-1} T_1 (B \vdots I). \qquad (4.46)$$

From which it follows $T_m T_{m-1} \cdots T_1 B = I$

$$\text{and } T_m T_{m-1} \cdots T_1 I = B^{-1} \qquad (4.47)$$

The solution value is of course given by

$$\beta^m = T_m T_{m-1} T_1 \beta. \qquad (4.48)$$

The elementary transformation matrices $T_m, T_{m-1}, \ldots, T_1$ may for all practical purposes replace the explicit expression for the inverse matrix B^{-1}. These transformation matrices may be compactly stored as vectors and are often referred to as "eta-vectors".

The method of carrying out the steps of the revised simplex algorithm with the eta-vectors instead of the B^{-1} may now be discussed. There are two pertinent steps where B^{-1} is used; these are:

Step 1. *Obtaining π the pricing vector*

The relationship

$$\pi = e.B^{-1}$$

may be replaced by the relationship

$$\pi = e(T_m T_{m-1} \cdots T_1) \qquad (4.49)$$

which implies that the form vector e should premultiply the series of transformation matrices $T_1, T_2, \ldots, T_m$ in the reverse order. This operation is known as "backward transformation".

Step 3. *Transform column*

The relationship for updating the chosen column q

$$\bar{a}_q = B^{-1} a_q$$

may be replaced by the relationship

$$\bar{a}_q = T_m T_{m-1} \ldots T_1 a_q; \tag{4.50}$$

this operation is known as "Forward Transformation".

To illustrate the principle of the method the iteration 4 of the example in Section 4.3 may be worked out using the product form algorithm. Applying Backward Transformation, the pricing vector is obtained.

$$\pi^4 = e . T_3 . T_2 . T_1 \tag{4.51}$$

$$= (1, 0, 0, 0)$$

$$\times \begin{bmatrix} 1 & 65/31 & & \\ & 29/31 & & \\ & -9/31 & 1 & \\ & -4/31 & & 1 \end{bmatrix} \times \begin{bmatrix} 1 & & & 4/29 \\ & 1 & & -26/29 \\ & & 1 & -1/29 \\ & & & 6/29 \end{bmatrix} \times \begin{bmatrix} 1 & & 1/3 & \\ & 1 & -2/3 & \\ & & 1/6 & \\ & & -1/6 & 1 \end{bmatrix}$$

$$= (1, 65/32, -24/31, -54/31)$$

The pricing operation then is the same as that illustrated in that section. After selecting the column a_4 corresponding to the variable x_4, this may be updated by the forward transformation.

$$\bar{a}_4 = T_3 T_2 T_1 a_4 \tag{4.52}$$

$$\begin{bmatrix} 1 & 65/31 & & \\ & 29/31 & & \\ & -9/31 & 1 & \\ & -4/31 & & 1 \end{bmatrix} \times \begin{bmatrix} 1 & & & 4/29 \\ & 1 & & -26/29 \\ & & 1 & -1/29 \\ & & & 6/29 \end{bmatrix} \times \begin{bmatrix} 1 & & 1/3 & \\ & 1 & -2/3 & \\ & & 1/6 & \\ & & -1/6 & 1 \end{bmatrix} \times \begin{bmatrix} -2 \\ 4 \\ 5 \\ 2 \end{bmatrix}$$

$$= \begin{bmatrix} -30/31 \\ -11/31 \\ 28/31 \\ 9/31 \end{bmatrix}$$

The other operations are then carried out as outlined in that section.

In the absence of large illustrative examples of LP problems processed by computer programs employing "product form algorithm" it is very difficult to convince a newcomer to the field, of the advantages accruing from this method. If this point is not appreciated then product form seems to make heavy weather of a very straight forward computation. The main motivation in representing the inverse of a matrix in this context is to make it as sparse as possible. Indeed one of the unsolved problems in numerical analysis is to represent the inverse of any matrix with the minimum number of non-zero elements.

To appreciate the power of the product form representation consider an example due to W. Orchard-Hays [43]. Given the matrix B; its inverse may be obtained by the sequence of pivotal steps

$$B = \begin{bmatrix} \textcircled{1} & & & & 1 \\ 1 & 1 & & & \\ & 1 & 1 & & \\ & & 1 & 1 & \\ & & & 1 & 1 \end{bmatrix} \to \begin{bmatrix} 1 & & & & 1 \\ -1 & \textcircled{1} & & & -1 \\ & 1 & 1 & & \\ & & 1 & 1 & \\ & & & 1 & 1 \end{bmatrix}$$

$$\to \begin{bmatrix} 1 & & & & 1 \\ -1 & 1 & & & -1 \\ 1 & -1 & \textcircled{1} & & 1 \\ & & 1 & 1 & \\ & & & 1 & 1 \end{bmatrix} \to \begin{bmatrix} 1 & & & & 1 \\ -1 & 1 & & & -1 \\ 1 & -1 & 1 & & 1 \\ -1 & 1 & -1 & \textcircled{1} & -1 \\ & & & 1 & 1 \end{bmatrix} \to$$

[*continued overleaf*]

[*continued from previous page*]

$$\rightarrow \begin{bmatrix} 1 & & & & 1 \\ -1 & 1 & & & -1 \\ 1 & -1 & 1 & & 1 \\ -1 & 1 & -1 & 1 & -1 \\ 1 & -1 & 1 & -1 & \textcircled{2} \end{bmatrix} \rightarrow \begin{bmatrix} \frac{1}{2} & \frac{1}{2} & -\frac{1}{2} & \frac{1}{2} & -\frac{1}{2} \\ -\frac{1}{2} & \frac{1}{2} & \frac{1}{2} & -\frac{1}{2} & \frac{1}{2} \\ \frac{1}{2} & -\frac{1}{2} & \frac{1}{2} & \frac{1}{2} & -\frac{1}{2} \\ -\frac{1}{2} & \frac{1}{2} & -\frac{1}{2} & \frac{1}{2} & \frac{1}{2} \\ \frac{1}{2} & -\frac{1}{2} & \frac{1}{2} & -\frac{1}{2} & \frac{1}{2} \end{bmatrix} \quad (4.53)$$

Observe that the explicit inverse is a full 5×5 matrix. In this sequence if one only noted the transformation matrices then these will be (storing only in the form of eta vectors) as shown in (4.54)

T_5	T_4	T_3	T_2	T_1
$-\frac{1}{2}$				$\textcircled{1}$
$\frac{1}{2}$			$\textcircled{1}$	-1
$-\frac{1}{2}$		$\textcircled{1}$	-1	
$\frac{1}{2}$	$\textcircled{1}$	-1		
$\textcircled{\frac{1}{2}}$	-1			

(4.54)

Pivotal elements are encircled: note that a total of 13 non zero elements represent the inverse. It is also worth pointing out at this stage how the choice and the order of pivot elements affect the build up of the nonzero elements. If for instance one pivoted backward up the leading diagonal leads to the sequence of transformation matrices as shown in (4.55).

T_5	T_4	T_3	T_2	T_1
$\textcircled{\frac{1}{2}}$	1	-1	1	-1
$-\frac{1}{2}$	$\textcircled{1}$			
$\frac{1}{2}$	-1	$\textcircled{1}$		
$-\frac{1}{2}$	1	-1	$\textcircled{1}$	
$\frac{1}{2}$	-1	1	-1	$\textcircled{1}$

(4.55)

This time 19 nonzero elements are required for a representation of the inverse of the matrix. This also explains why in LP codes from time to time while carrying out iterations and creating eta vectors, this process is suspended and a "reinversion" procedure is carried out. This consists of finding out within known strategy a very sparse representation of the inverse of the current basis. It goes without saying that the sparser this representation is the quicker is the process of backward and forward transformations and hence the iteration speed is higher. Fortunately enough this also improves the numerical accuracy as less numerical operations are carried out to obtain these transformations. This topic is again considered in Chapter 5, Section 4.

EXERCISES

4.1 Solve the linear programming problem

$$\begin{aligned} \text{maximize} \quad & z = 3x_1 + 2x_2 + 2x_3 \\ \text{subject to} \quad & 3x_1 + 4x_2 + 4x_3 \leqslant 5 \\ & 2x_1 + 6x_2 + 5x_3 \geqslant 6 \\ & x_1 + x_2 + 2x_3 = 2 \\ \text{and} \quad & x_1, x_2, x_3 \geqslant 0. \end{aligned}$$

Express the inverse of the optimal basis matrix as the product of three elementary transformation matrices. Carry out in the revised simplex method the final iteration by which one establishes the optimality of the final solution.

4.2 State the five steps which occur in a typical iteration of the product form of the inverse matrix method for the solution of ordinary linear programming problems. Obtain the inverse of the matrix

$$\begin{bmatrix} 4 & 1 & 1 & 1 \\ 1 & 1 & & \\ 1 & & 1 & \\ 1 & & & 1 \end{bmatrix}$$

as a product of elementary transformation matrices. Is the product form representation of the inverse of a matrix unique? Can you find a second set of transformation matrices the product of which also represents the inverse of the above matrix?

4.3 Discuss the role of B^{-1} the inverse of the current basis matrix, in the inverse matrix (revised simplex) method of linear programming. Use this method to solve the problem

$$\begin{aligned} \text{Maximize} \quad & z = 4x_1 - x_2 + x_3 - 10x_4 + 6x_5 \\ \text{subject to} \quad & x_1 - x_2 - x_3 + x_4 = 1 \\ & 2x_1 + x_2 + 2x_3 + x_5 = 7 \\ & 3x_1 - x_2 - x_3 = 4 \\ \text{and} \quad & x_1, x_2 \ldots x_5 \geqslant 0. \end{aligned}$$

If the right-hand side vector is changed from $(1, 7, 4)^T$ to $(2, 6, 5)^T$ does the optimal basis continue to be feasible?

CHAPTER 5

Computational Refinements and Extensions within the Context of the Revised Simplex Method

KEYWORDS: Multiple Pricing, Composite Simplex Algorithm, Simple Upper Bound, Generalised Upper Bound, LU Decomposition, Elimination Form of Inverse.

5.1 COMPUTER IMPLEMENTATION OF THE PRODUCT FORM OF INVERSE: SOME RELATED TOPICS

Linear programming systems as implemented on present day computers provide excellent examples of how large amounts of structured data should be manipulated. Various schemes of packing data to represent large sparse matrices are in vogue. In general if a problem with m rows and $m + n$ columns is represented then only $m(m + n) \times \rho_x$ real quantities and their suitably packed indices need be stored; $\rho_x(0 < \rho_x \leqslant 1)$ is of course the sparsity of the original constraint matrix. Let ρ_T denote the sparsity of the tableau from iteration to iteration. In order to demonstrate the advantage of the inverse matrix method over the extended tableau method, it is convenient to assume that $\rho_x = \rho_T = 1$. In this case the former method requires $m \times m$ elements and the latter requires $m \times n$ elements to be updated in one simple step. For problems with many more number of columns than rows ($n \gg m$) and this is often the case, the corresponding saving in elementary operations takes place.

To demonstrate the advantage of the product form algorithm the sparsity should be taken into consideration; also note that an $m \times m$ square matrix is not updated in every iteration. The pricing vector π is normally full and the pricing of the matrix therefore requires $m \times n \times \rho_x$ multiplication†

† In computers multiplication time is an order of magnitude higher than that of addition and subtraction; hence only multiplication is taken as the basic unit to quantify computing effort.

operations. The updating of the tableau, however, requires $m \times n \times \rho_T^2$ multiplication operations.

This would itself be a forceful argument for using the tableau method if ρ_T^2 was smaller than ρ_x. In real life problems the contrary is true, viz., the nonzeros of the tableaus are far more numerous than the nonzeros of the initial constraint matrix and ρ_T^2 is greater than ρ_x: the example in Chapter 4, Section 4 goes towards subtantiating this claim.

The product form algorithm and its method of implementation, however, leads to a few other natural advantages.

(i) The whole LP matrix need not be "priced" in one pass.

(ii) Depending on the number of full vector areas available, i.e., number of m locations to store real quantities, that many number of candidate columns may be updated, transformed and reconsidered before repricing the matrix. This is a process known as "multiple pricing".

(iii) After some number of iterations, the product form of inverse (PFI) representation gets too full and a reinversion may be carried out. This creates a new set of eta vectors by a reordered sequence of pivot steps to make the representation sparse.

The difficulty of solving an LP problem therefore in the first place depends on m, i.e., the number of rows and hence the number of regions available, a region is where the explicit and fully updated column vectors are manipulated by direct access. In the second place the initial sparsity of the matrix determines the speed of pricing operation and forward transformation hence the iteration speed. The sparse product form obtained after reinversion produces two beneficial effects:

(a) reduces processing speed for forward and backward transformation.

(b) reduces rounding errors.

A brief outline of a typical LP system and its facilities is provided in Appendix 2.

5.2 COMPOSITE SIMPLEX ALGORITHM AND THE PHASE 1 CONVERGENCE

The Phase 1 pivot choice rules outlined in Sections 2 and 3 of Chapter 3 guarantee convergence but also admit of many refinements and improvements. In the early days it was considered that a combination of Phase 1 and Phase 2 objectives is a sensible form to work with. This approach, to start with, complicates the test for "no feasible" condition; it is also not very meaningful without suitable weighting of the two objective forms. Another concept in these lines is the logical pricing strategy; in this a column with negative d_j for both Phase 1 and Phase 2 objective form is chosen. The objective form of the earlier method is often referred to in literature as the composite objective function. But most of the present day LP systems do not necessarily provide these facilities; instead these employ sophisticated pivot row choice rules which may remove multiple infeasibilities in one pivot step. This approach and its variations are well described by Wolfe [56]. This leads to different row choice rules in the two phases of the simplex method but it otherwise fits in very well with the steps of the simplex method. Because of historical association with the previous method this method is called the "Composite Simplex Method".

Before going into the details of the method a few words on the general aims will be in order. Let,

$$N_f = \sum_{i=1}^{m} \delta_i, \quad \text{where} \quad \begin{aligned} \delta_i &= 1 \text{ if } i \in A_t \cup N_g \;\dagger \\ &= 0 \text{ otherwise} \end{aligned} \tag{5.1}$$

A_t and N_g are as defined in Section 3, Chapter 3. The sum of infeasibilities is defined as

$$S = \sum_{i \in N_g} \beta_i - \sum_{i \in A_t} \beta_i. \tag{5.2}$$

N_f is referred to as the number of infeasibilities and S the sum of infeasibilities; it also follows that in a feasible solution both S and N_f vanish. From the viewpoint of Phase 1 convergence, however, N_f is a more important measure than S. Ability to reduce N_f by integral amount is often more desirable than the reduction of the magnitude of S. It usually happens that both N_f and S reduce together but it is indeed possible to reduce N_f and yet increase the value of S. The strategy therefore has been to work with the infeasibility form associated with S and use this to price vectors into the basis or to detect no feasible solutions during row choice; subsequently, one attempts to reduce N_f, i.e. the number of infeasibilities.

To illustrate this approach consider the Tableau 5.1 which is infeasible and in which all the y variables are the basic variables. Assume x_q is the variable priced to come into the basis. The real variable to leave the basis by the simple pivot choice rule of the last chapter is obtained by finding the minimum ratio,

$$\min_i \{\beta i/\bar{a}_{iq} | \beta_i \quad \text{and} \quad \bar{a}_{iq} \text{ of the same sign and } \bar{a}_{iq} \neq 0\}. \tag{5.3}$$

Let p be the row index for which this minimum is achieved. Fig. 5.1 illustrates the various alternative levels at which x_q may come into basis.

† An artificial variable in the basis at zero level may not be considered an infeasibility and thus may not belong to the set A_t.

	1	$-x_1$	$-x_2$	$-x_q$	$-x_n$
Z	β_0			$\bar{C}_q$	
y_1	β_1			$\bar{a}_{1q}$	
y_2	β_2				
y_p	β_p			$\bar{a}_{pq}$	
y_m	β_m			$\bar{a}_{mq}$	

TABLEAU 5.1.

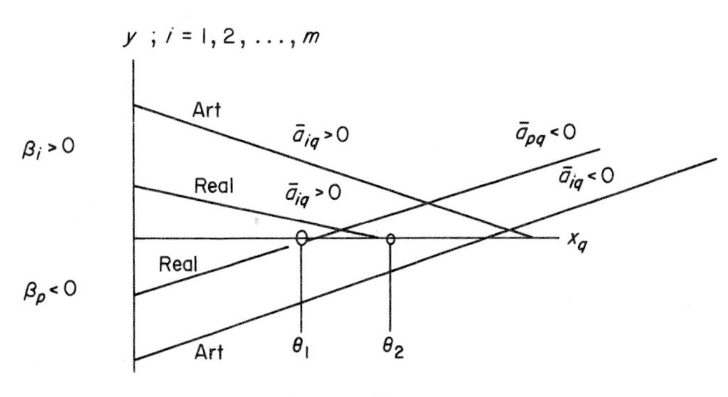

FIGURE 5.1

By this rule x_q comes into the basis at $\beta_p/\bar{a}_{pq}$. By the modified rule, however, the variable x_q is increased until one of the following occurs:

(a) A genuine artificial variable vanishes, $\Big\}\ \theta_2$
(b) A variable that is now positive becomes zero
(c) All the negative variables which may be affected $a_{iq} < 0$ are made non-negative θ_1,
(d) An artificial variable at zero level is removed.

To state these rules in mathematical terms define the following sets of variable indices.

N_r: The set of row indices of the slack or structural variables at negative level.
A_t: The set of row indices of the artificial variables at positive level.
A_{-t}: The set of row indices of the artificial variables at negative level.
I_p: The set of row indices of structural or slack variables at positive level.
N_{ze}: The set of row indices of structural or slack variables at a level

$$-z_{\text{tolb}} \leqslant \beta_i \leqslant z_{\text{tolb}}$$

A_{ze}: The set of row indices of artifical variables at a level $z_{\text{tolb}} \leqslant \beta_i \leqslant z_{\text{tolb}}$. z_{tolb} is a tolerance used to determine the solution values at zero level and not count artificials as infeasibilities. Also observe that $N_g = N_r \cup A_{-t}$.

Four alternative row choice possibilities may be now defined.

Case 1. Choose as the variable to leave the basis, the last non-artifical variable to rise to zero from a negative value

$$\theta_1 = \max_{i \in N_r} \{\beta_i/\bar{a}_{iq} | \bar{a}_{iq} < 0\}. \tag{5.4}$$

Case 2. Choose as the variable to leave the basis the first non-artificial variable to drop to zero or the first artificial variable to reach zero (giving priority to artificials).

$$\theta_2 = \min_{i \in I_p \cup A_t \cup A_{-t}} \{\beta_i/\bar{a}_{iq} | \beta_i, \bar{a}_{iq} \text{ same sign and } \bar{a}_{iq} \neq 0\}. \tag{5.5}$$

Case 3. Choose as the variable to leave the basis the virtually zero non-artificial variable having the largest positive pivot element.

$$\theta_3 = \frac{\beta_p}{\bar{a}_{pq}}, \text{ where } \bar{a}_{pq} = \max_{i \in N_{ze}} \{\bar{a}_{iq} | \bar{a}_{iq} > 0\}. \tag{5.6}$$

Case 4. Choose as the variable to leave the basis the artificial variable with the largest absolute pivot corresponding to a zero solution value.

$$\theta_4 = \frac{\beta_p}{\bar{a}_{pq}}, \text{ where } \bar{a}_{pq} = \max_{i \in A_{ze}} \{\text{abs}(\bar{a}_{iq}) | \bar{a}_{iq} \neq 0\}. \tag{5.7}$$

The pivot row is then chosen by the following rule: If Case 4 exists Cases 3, 2, 1 are not considered. If Case 3 exists Cases 2 and 1 are not considered. If $\theta_2 > \theta_1$ then Case 1 is not considered, if Case 2 does not exist Case 1 is used to obtain the variable which must leave the basis.

Case 4 and Case 3 are important in computer codes designed to solve sizeable problems which are often degenerate at the time of formulation. The advantage of using the two cases, Case 1 and Case 2 may be demonstrated by the following illustrate example.

$$\begin{aligned} \text{Minimize} \quad & f(x) = 6x_1 + 7x_2 + 7x_3 + 8x_4 \\ \text{subject to} \quad & x_1/2 + x_2 + x_3 + x_4 = 1 \\ & 6x_2 + 3x_3 + 6x_4 \geqslant 3 \\ & 3x_2 + 3x_3 + 3x_4 \geqslant 2 \\ & 12x_1 + 4x_3 + 2x_4 = 6 \end{aligned}$$

As outlined in Section 3, Chapter 3, for a compact method only two artificial variables x_{t1} and x_{t4} need to be introduced for the two equalities and the variables x_5 and x_6 are introduced as two slacks starting at negative levels. The problem is set up in this form in Tableau 5.2. The infeasibility form is

		$-x_1$	$-x_2$	$-x_3$	$-x_4$
$f(x)$	0	6	7	7	8
x_{t1}	1	$\frac{1}{2}$	1	1	1
x_5	−3		−6	−3	−6
x_6	−2		−3	−3	−3
x_{t4}	6	(12)		4	2

TABLEAU 5.2

Tableau 5·8

		$-x_{t4}$	$-x_2$	$-x_3$	$-x_4$
$f(x)$	−3	$-\frac{1}{2}$	7	(5)	7
x_{t1}	$\frac{3}{4}$	−1/24	(1)	5/6	11/12
x_5	−3		(−6)	−3	−6
x_6	−2		−3	−3	−3
x_1	$\frac{1}{2}$	1/12		1/3	1/6

TABLEAU 5.3

		$-x_{t4}$	$-x_5$	$-x_3$	$-x_4$
$f(x)$	−13/2	$-\frac{1}{2}$	7/6	3/2	
x_{t1}	$\frac{1}{4}$	1/24	1/6	(1/3)	−1/12
x_2	$\frac{1}{2}$		−1/6	1/2	1
x_6	$-\frac{1}{2}$		$-\frac{1}{2}$	(−3/2)	
x_1	$\frac{1}{2}$	1/12		1/3	1/6

TABLEAU 5.4

		$-x_{t4}$	$-x_5$	$-x_6$	$-x_4$
$f(x)$	−7	$-\frac{1}{2}$	4/6	1	
x_{t1}	5/36	1/24	1/18	(2/9)	−1/12
x_2	1/3		−1/3	1/3	1
x_3	1/3		1/3	−2/3	
x_1	7/18	1/12	−1/9	2/9	1/6

TABLEAU 5.5

		$-x_{t4}$	$-x_5$	$-x_{t1}$	$-x_4$
$f(x)$	−61/8	−11/16	5/12	−9/2	3/8
x_6	5/8	3/16	1/4	9/2	−3/8
x_2	1/8	−1/16	−5/12	−3/2	9/8
x_3	3/4	1/8	1/2	3	−1/4
x_1	1/4	1/24	−1/6	1	1/4

TABLEAU 5.6

		$-x_{t4}$	$-x_5$	$-x_{t1}$	$-x_4$
$f(x)$	−61/8	−11/16	5/12	−9/2	3/8
x_3	3/4	1/8	1/2	3	−1/4
x_2	1/8	−1/16	−5/12	−3/2	9/8
x_6	5/8	3/16	1/4	9/2	−3/8
x_1	1/4	1/24	−1/16	1	1/4

TABLEAU 5.7

		$-x_{t4}$	$-x_{t1}$	$-x_3$	$-x_4$
$f(x)$	$-33/4$	$-5/24$	-7	$-5/6$	$7/12$
x_2	$3/4$	$-1/24$	1	$5/6$	$11/12$
x_5	$3/2$	$-1/4$	6	2	$-1/2$
x_6	$1/4$	$-1/8$	3	$-1/2$	$-1/4$
x_1	$1/2$	$1/12$		$1/3$	$1/6$

TABLEAU 5.8

obtained by the relationship (3.12) and (3.13) given in Chapter 3.

$-w(x) = -12 + 25/2(-x_1) - 10(-x_2) - 11(-x_3) - 12(-x_4)$ in Tableau 5.2

Applying the basic pivot choice rule (5.3) the rows corresponding to minimum non-negative ratio are chosen and in 4 steps the first feasible solution is obtained.

In Tableaux 5.3 and 5.4 the variations corresponding to Case 2 and Case 1 may be applied. In Tableau 5.3 $\theta_2 = 3/4 > \theta_1 = 2/3$ and the exchange $x_{t1} \leftrightarrow x_3$ leads to the feasible Tableau 5.8. In Tableau 5.4 again by the same criterion if the exchange $x_{t1} \leftrightarrow x_3$ is carried out the feasible Tableau 5.7 is obtained.

5.3 SIMPLE UPPER BOUND ALGORITHM

In the last ten years most significant computational advantages have been gained by incorporating "Simple Upper Bound" (SUB) and "Generalized Upper Bound" (GUB) algorithms in large-scale LP systems. The description of this and the next section closely follows that due to Beale [3] on the topic.

Under the normal LP computational scheme a lower bound on a variable,

$$l_j \leqslant x_j \tag{5.8}$$

does not pose any difficulty. One only needs to introduce a variable $w_j \geqslant 0$, and eliminate x_j from the tableau by substituting the relationship,

$$x_j = w_j + l_j. \tag{5.9}$$

Thus (5.8) does not introduce explicit rows in the problem. For an upper bound relation on a variable,

$$x_j \leqslant h_j \tag{5.10}$$

the bounding constraint can be written as

$$x_j + x_j' = h_j, \qquad x_j \geqslant 0. \tag{5.11}$$

Substitution of (5.11) does not ensure that in the program x_j or x'_j is less than h_j whereas the simplex method ensures the non-negativity of variables. Thus the substitution of (5.9) is only natural. Hence, inequalities of the form (5.10) must be explicitly introduced.

In an upper bound algorithm, however, this constraint (5.10) can be implicitly taken into account; towards this it is convenient to introduce a non-positive variable $y_j \leqslant 0$, such that

$$x_j = y_j + h_j \tag{5.12}$$

and modify the simplex algorithm to ensure that $-h_j \leqslant y_j \leqslant 0$ holds and that y_j and x_j when they are non basic are priced suitably.

The modifications of the simplex algorithm to take into account these bounded variables are as follows. Only the changes to be added in each step of the algorithm stated in Section 4.2 are noted.

Step 1

While setting up the form vector e set $e_i = 1$ if the variable x_{p_i} in the ith row exceeds its upper bound h_{p_i}.

Step 2

If a non-basic variable x_q is out of the basis at its upper bound then reverse the sign of the reduced cost coefficient of this variable since one is interested only in the non-positive values of the variable y_q.

Step 3

Remains completely unaltered.

Step 4

Assume that x_q is the non basic variable chosen to come into the basis and it is at its lower bound.

Define R_1 as the set of rows such that $\bar{a}_{i0} \geqslant 0$ and $\bar{a}_{iq} > 0$ or alternatively $\bar{a}_{i0} \leqslant 0$ and $\bar{a}_{iq} < 0$ and let the variable pivoted in row i be an artificial; define p_1 and θ_1 such that

$$\frac{\bar{a}_{p_1 0}}{\bar{a}_{p_1 q}} = \min_{i, i \in R_1} \frac{\bar{a}_{i0}}{\bar{a}_{iq}} = \theta_1 \, . \tag{5.13}$$

Define R_2 as the set of rows such that $\bar{a}_{i0} \leqslant 0$, and $\bar{a}_{iq} < 0$ and x_{p_i} is neither artificial nor bounded; and define p_2, θ_2 such that

$$\frac{\bar{a}_{p20}}{\bar{a}_{p2q}} = \max_{i,\, i \in R_2} \frac{\bar{a}_{i0}}{\bar{a}_{iq}} = \theta_2 \tag{5.14}$$

Let R_3 be the set of rows such that $\bar{a}_{i0} < h_{r_i}$, $\bar{a}_{iq} < 0$ and the bounded variable x_{r_i} is pivoted in the i th row; define p_3 and θ_3 such that

$$\frac{\bar{a}_{p30} - h_{p3}}{\bar{a}_{p3q}} = \min_{i, i \in R_3} \frac{\bar{a}_{i0} - h_{r_i}}{\bar{a}_{iq}} = \theta_3. \tag{5.15}$$

If x_q is a bounded variable out of the basis at its upper or lower bound then define $\theta_4 = h_q$ otherwise define $\theta_4 = \infty$.
Then obtain

$$\theta = \theta_k = \min(\theta_1, \theta_3, \theta_4). \tag{5.16}$$

unless these are all equal to ∞ or undefined in which case $\theta = \theta_2$ and $k = 2$. If θ_1 or θ_2 is chosen pivot in the row p_1 or p_2. If θ_3 is chosen, replace a_{p30} by $a_{p_30} - h_{p_3}$ and note that x_{p_3} is now out of the basis at its upper bound. For $\theta = \theta_4$ no pivotal operation is required x_q moves from the lower to its upper bound; this needs to be recorded and the solution values are updated by the relationship

$$\bar{a}'_{i0} = \bar{a}_{i0} - \bar{a}_{iq} \times h_q, \qquad i = 0, 1, 2, \ldots, m. \tag{5.17}$$

If the variable x_q is at its upper bound at the beginning of the iteration, the procedure is more or less the same except that all the tests are performed on $-\bar{a}_{iq}$ rather than $\bar{a}_{iq}$, after a pivotal step. However, h_q must be added to the value of x_q now brought to the basis. In all cases record that x_q is no longer at its upper bound.

The use of upper bounds also modifies the solution update procedure. This now takes the form $B^{-1}\bar{b}$, where $\bar{b}$ is the modified right hand side given by the relationship

$$\bar{b}_i = b_i - \sum_{j \in s} a_{ij} h_j \tag{5.18}$$

where s is the set of indices of the variables which are non basic and also at their upper bounds.

5.4 GENERALIZED UPPER BOUNDS.

The simple upper bound algorithm is computationally desirable since one may work with a contracted basis. A large number of real life problems exhibit mutually exclusive row structure made up of summation rows. These rows can be directly eliminated and one may work with a contracted basis Dantzig and Van-Slyke[16] first devised an algorithm taking advantage of the generalized upper bounding structure of these rows. It has since become a standard feature of the present day LP systems.

Consider the problem

$$\text{Maximize } x_0,$$

$$\text{Subject to } x_0 + \sum_j a_{0j}x_j + \sum_k \sum_j a_{0jk}x_{jk} = b_0$$

$$\sum_j a_{ij}x_j + \sum_k \sum_j a_{ijk}x_{jk} = b_i \qquad (i = 1, 2, \ldots, m) \qquad (5.19)$$

$$\sum_j x_{jk} = b_{m+k} \qquad (k = 1, 2, \ldots, t)$$

$$x_{jk} \geqslant 0 \text{ for all } j, k.$$

The last t rows are often referred to as convexity rows, or GUB rows and the variables occurring in such a row may be referred to as GUB set variables. Note that no GUB set variable may appear in more than one GUB row.

A very large percentage of the LP problems which are formulated and solved fall in this category; the problems which exhibit such structure are considered also in Chapter 8 which is concerned explicitly with the formulation of applications models.

In order to express the relationship of (5.19) more compactly let the variables in a non-GUBset be expressed with a second subscript of zero. The equation set may be then expressed as

$$x_0 + \sum_{k=0}^{t} \sum_j a_{0jk}x_{jk} = b_0$$

$$\sum_{k=0}^{t} \sum_j a_{ijk}x_{jk} = b_i \qquad (i = 1, 2, \ldots, m) \qquad (5.20)$$

$$\sum_j x_{jk} = b_{m+k} \qquad (k = 1, 2, \ldots, t).$$

Any basic solution to this problem must have at least one basic variable from each GUB set. So one such variable, say $x_{j_k k}$ can be selected and called the key variable for the set.

The key variables in the ordinary rows may be eliminated using the GUBrows and (5.20) may be rewritten as

$$x_0 + \sum_{k=0}^{t} \sum_j a_{0jk}x_{jk} = b_0* \qquad (5.21)$$

$$\sum_j \sum_k a^*_{ijk}x_{jk} = b^*_i \qquad (i = 1, 2, \ldots, m)$$

where

$$\left.\begin{aligned} a^*_{ijk} &= a_{ijk} - a_{ij_kk} \\ b^*_i &= b_i - \sum_{k=1}^{t} a_{ij_kk}\, b_{m+k} \end{aligned}\right\} \quad (i = 0, 1, 2, \ldots, m).$$

For these formulae to apply for $k = 0$ it is necessary to define $a_{ij_00} = 0$ for all i.

Considered now a basic solution to the shortened form of the problem expressed in (5.21) and let $x_0, x_{r_1}, x_{r_2}, \ldots, x_{r_m}$ be the set of basic variables. Assume that (5.21) is premultiplied by the inverse of the basic matrix of the contracted problem. The following form is then obtained

$$x_0 = \bar{a}_{00} + \sum_{k=1}^{t} \sum_j \bar{a}_{0jk}(-x_{jk})$$

$$x_{r_i} = \bar{a}_{i0} + \sum_{k=1}^{t} \sum_j \bar{a}_{ijk}(-x_{jk}) \qquad (i = 1, 2, \ldots, m). \tag{5.22}$$

This may be regarded as the top half of a contracted tableau; the variables on the right-hand side do not include the key variables or the basic variables. Given the relationship (5.22) the bottom half of the tableau may be constructed by substituting (5.22) in the last t equations of (5.20):

$$\left.\begin{aligned} x_{j_ll} &= \bar{a}_{m+l,0} + \sum_k \sum_j \bar{a}_{m+l,\,jk}(-x_{jk}) \\ \text{where } \bar{a}_{m+l,0} &= b_{m+l} - \sum_{i\in s_l} \bar{a}_{i0} \\ \bar{a}_{m+l,\,jk} &= \delta_{kl} - \sum_{i\in s_l} \bar{a}_{ijk} \end{aligned}\right\}, \quad (l = 1, 2 \ldots t), \tag{5.23}$$

where $\delta_{kl} = 1$ for $k = l$ and $\delta_{kl} = 0$ otherwise and s_l denote the set of indices of those rows in which variables from the lth set appear in the basis.

The relationships (5.22) and (5.23) may be used to reconstruct the entire contracted tableau of the original problem and all the necessary simplex computational steps may be carried out using a compact basis m long whereas there are $m + t$ constraints in the original problem. For further computational details refer to [3].

5.5 THE ELIMINATION FORM OF THE INVERSE AND THE ADVANTAGE OF ITS USE

The material in this section is based on the very recent developments in the compact representation of the inverse of a basis matrix and the method of completely updating such inverses [4] [10] [47] [48]. The presentation follows closely the recent developments in this field, it also provides a chance to consider briefly the mystical process of "reinversion" [43].

In real life LP problems the basis matrix is invariably sparse and by a suitable algorithm [43] it can be always partitioned as

$$B = \begin{bmatrix} \Lambda_1 & & \\ A & P & \\ D & E & \Lambda_2 \end{bmatrix}, \tag{5.24}$$

where Λ_1 and Λ_2 are lower triangular matrices, A, D, P, E are non-triangular matrices with defined boundaries; but these matrices are not necessarily full. The matrix P is often simply referred to as the "bump".

The matrix B may be factored as

$$B = \begin{bmatrix} \Lambda_1 & & \\ A & I & \\ D & & I \end{bmatrix} \times \begin{bmatrix} I & & \\ & P & \\ & & I \end{bmatrix} \times \begin{bmatrix} I & & \\ & I & \\ & E & I \end{bmatrix} \times \begin{bmatrix} I & & \\ & I & \\ & & \Lambda_2 \end{bmatrix}.$$

Using the diagonal terms as pivots eta vectors may be produced for the first, third and the last of these matrices and the number of elements so created are no more than those in the original matrix. The product form representation the P^{-1} is, however, nontrivial and by and large more dense than P.

It is well established with the numerical analysts that the inversion via LU decomposition which follows from Gaussian elimination is an attractive numerical tool [23]. By this approach B may be factorized as

$$B = LU, \tag{5.26}$$

and the inverse expressed as

$$B^{-1} = U^{-1}L^{-1}.$$

Now invoking product form representation of U^{-1} and L^{-1} the resulting inverse may be expressed in terms of $2m$ eta vectors

$$B^{-1} = U_1^{-1}U_2^{-1} \dots U_m^{-1} L_m^{-1} L_{m-1}^{-1} \dots L_1^{-1}. \tag{5.27}$$

Although it is not apparent from the partitioning and factorization of (5.24) and (5.25) the reinversion process in most of the modern LP systems produce exactly such a decomposition. Beale [4] therefore suggests the following interpretation of the reinversion process. Express the partition of B as

$$B = \begin{bmatrix} \Lambda_1 & & \\ {}_rD & {}_{cr}\Lambda_2 & {}_rE \\ A & & P \end{bmatrix} \tag{5.28}$$

where the last two column partitions of B are permuted and the last two row partitions are also permuted from the representation in (5.24). Further the rows of ${}_rD$, ${}_rE$ are set out in reverse order to those of D, E, and the rows and columns of ${}_{cr}\Lambda_2$ are in reverse order to that Λ_2. If Gaussian elimination is now carried out on this matrix the first part, pivoting down the diagonal of Λ_1 yields eta vectors corresponding to L_1^{-1}, L_2^{-1} and the second partition U_m^{-1}, U_{m-1}^{-1} and factorization of P adds to both L^{-1} and U^{-1}.

This inverse representation therefore for obvious reasons is termed the Elimination Form of Inverse or EFI. The EFI is used not simply because of its numerical curiosity or the sparsity of the resulting inverse. It so happens that it facilitates the compact updating of the neighbouring bases. If these ideas [10] concerning sparse updating are cleverly implemented these lead to considerable computational efficiency. This topic is fully discussed in [10], [40], [47], [48].

EXERCISES

5.1. Discuss the role of reinversion in a linear programming code. Obtain the Elimination Form of the Inverse (EFI) of the two matrices discussed in the illustration and the exercise set out in chapter 4.

5.2. Apply the composite simplex algorithm as discussed in this chapter to the problem in Exercise 2.5.

5.3. Discuss the advantages of using the simple Upper Bound and the Generalized Upper Bound algorithms.

CHAPTER 6

Duality Properties of Linear Programs and Post Optimal Analysis

KEYWORDS: Primal Problem, Dual Problem, Skew Symmetry, Complementary Slackness, Orthogonality Condition, Cutting Planes, Cost Ranging, RHS Ranging, RHS Parametrics, Objective Parametrics, Composite Parametrics.

6.1 DUALITY OF MATHEMATICAL SYSTEMS AND THE SYMMETRY PROPERTY OF THE TRANSFORMATION RULES

Some mathematical systems are found to possess an esoteric property: given the original system (say of equations) there exists an associated system made up of coefficients which are the descriptors of the original system, such that this second system possesses (reflexiv) symmetry features in relation to the original system. This symmetry is known as the duality of the system and this property may be exploited as an alternative means of solving or analysing the original system.

In electrical circuit theory, for instance, current generator circuits and voltage generator circuits display a duality relationship: a relationship often made use of by adopting the most convenient of the two representations for solving the circuit theory problems.

Returning to the system of linear equations it is observed that there exists a skew symmetry property associated with the pivotal transformation that is carried out to solve such a system. Before this is described note that the solvability of a system of equations and its transpose is discussed in Appendix 1.1 and the Minkowski–Farkas Lemma concerning the solvability of such systems in terms of non-negative variables is also presented in this appendix.

Consider the system of linear equations

$$x_0 = \bar{a}_{00} + \sum_{j=1}^{n} \bar{a}_{0j}(-x_j),$$

$$x_{n+i} = \bar{a}_{i0} + \sum_{j=1}^{n} \bar{a}_{ij}(-x_j) \qquad (i = 1, 2, \ldots, m,) \tag{6.1}$$

and the system of equations made up of the transposed coefficients of the first system viz

$$u_0 = \bar{a}_{00} + \sum_{i=1}^{m} \bar{a}_{i0} u_{n+i}$$

$$u_j = \bar{a}_{0j} + \sum_{i=1}^{m} \bar{a}_{ij} u_{n+i} \qquad (j = 1, 2, \ldots, n.) \tag{6.2}$$

If the qth row in the second system of equation is pivoted against the pth column i.e. the variable u_{n+p} is expressed in terms of $u_{n+1}, u_{n+2}, \ldots, u_{n+m}$ and u_q, then the transformation rules for this set of equations may be derived as

$$\begin{aligned}
\bar{a}'_{pq} &= 1/\bar{a}_{pq} \\
\bar{a}'_{pj} &= \bar{a}_{pj}/\bar{a}_{pq} \qquad j = 0, 1, 2, \ldots, n, \text{ and } j \neq q \\
\bar{a}'_{iq} &= -\bar{a}_{iq}/\bar{a}_{pq} \qquad i = 0, 1, 2, \ldots, m, \text{ and } i \neq p \\
\bar{a}'_{ij} &= \bar{a}_{ij} - (\bar{a}_{iq} \cdot \bar{a}_{pj})/\bar{a}_{pq}, \text{ for all } i, j \text{ and } i \neq p \text{ or } j \neq q.
\end{aligned} \tag{6.3}$$

It may now be observed that these are exactly the transformation rules derived in Chapter 2, for the set of equations (6.1) when the pth row is exchanged with the qth column; note that in the second system the variables are not prefixed by a negative sign. The transformation rules may be looked upon as possessing a "Skew-Symmetry" property since the column and row transformation relations are the same except for a negative sign.

6.2 DUALITY OF THE CANONICAL FORM OF LINEAR PROGRAM

Consider the linear programming problem stated in the canonical form

$$A_c\text{: Maximize } f(x) = c_1 x_1 + c_2 x_2 + \ldots + c_n x_n$$

$$\begin{aligned}
\text{subject to} \qquad & a_{11}x_1 + a_{12}x_2 + \ldots + a_{1n}x_n \leqslant b_1 \\
& a_{22}x_1 + a_{22}x_2 + \ldots + a_{2n}x_n \leqslant b_2 \\
& a_{m1}x_1 + a_{m2}x_2 + \ldots + a_{mn}x_n \leqslant b_m \\
\text{and} \qquad & x_1, x_2, \ldots x_n \geqslant 0,
\end{aligned} \tag{6.4}$$

or in matrix notation,

$$\text{Maximize} \qquad f(x) = c'x,$$
$$\text{subject to} \qquad Ax \leqslant b, \text{ and } x \geqslant 0.$$

Consider further the associated problem,

$$
\begin{aligned}
B_c\text{: Minimize} \qquad & G(v) = b_1v_1 + b_2v_2 + \ldots + b_mv_m, \\
\text{subject to} \qquad & a_{11}v_1 + a_{21}v_2 + \ldots + a_{m1}v_m \geqslant c_1, \\
& a_{12}v_1 + a_{22}v_2 + \ldots + a_{m2}v_m \geqslant c_2, \\
& \cdots \\
& a_{1n}v_1 + a_{2n}v_2 + \ldots + a_{mn}v_m \geqslant c_m, \\
\text{and} \qquad & v_1, v_2, \ldots, v_m \geqslant 0,
\end{aligned} \tag{6.5}
$$

or in matrix notation

$$\text{Minimize } G(v) = b'v,$$
$$\text{subject to } A'v \geqslant c, \text{ and } v \geqslant 0.$$

If the problem A_c is called the "Primal Problem" and the components of x the primal variables in which the problem is solved then B_c is called the "Dual Problem" and the components of v the "dual variables". This terminology can be obviously interchanged. Introduce m slack variables $y_i \geqslant 0$ ($i = 1, 2, \ldots, m$) for the problem A_c and a typical equation of the primal problem becomes

$$a_{i1}x_1 + a_{i2}x_2 + \ldots + a_{in}x_n + y_i = b_i. \tag{6.6}$$

similarly introduce n surplus variables $u_j \geqslant 0$, ($j = 1, 2, \ldots, n$) for the problem B_c and a typical equation of the dual problem becomes

$$a_{1j}v_1 + a_{2j}v_2 + \ldots + a_{mj}v_m - u_j = c_j. \tag{6.7}$$

It therefore follows that the Tableau 6.1 can represent the primal and the dual problem at the same time, if only it is observed that the equation of the

	$G(v)$	$-x_1$	$-x_2$	$\ldots$	$-x_n$	
$f(x)$		$-c_1$	$-c_2$		$-c_n$	
x_{n+1} or y_1	b_1	a_{11}	a_{12}	$\ldots$	a_{1n}	v_1 or u_{n+1}
x_{n+2} or y_2	b_2	a_{21}	a_{22}	$\ldots$	a_{2n}	v_2 or u_{n+2}
$\vdots$	$\vdots$	$\vdots$	$\vdots$		$\vdots$	
x_{n+m} or y_m	b_m	a_{m1}	a_{m2}	$\ldots$	a_{mn}	v_m or u_{n+m}
		u_1	u_2		u_n	

TABLEAU 6.1

type (6.6) are set up row wise and those for (6.7) are set up column wise. Also note in passing that the primal and dual variables may be redefined

y_1 as x_{n+1}, y_2 as $x_{n+2}, \ldots, y_m$ as x_{n+m}; v_1 as u_{n+1}, v_2 as $u_{n+2}, \ldots, v_m$ as u_{n+m};

the advantage of doing this becomes apparent later on in this section.

The following properties of the problem pair may now be noted (for detailed theoretical discussions see [1], [26], [45]). (a) If feasible solutions exist to both A_c and B_c then the cost of any feasible solution to A_c must be less than or equal to the cost of any feasible solution to B_c i.e.,

$$f(\bar{x}) \leqslant G(\bar{v}). \tag{6.8}$$

further, the optimum solution to both the problems is attained when the equality,

$$f(x^0) = G(v^0), \tag{6.9}$$

holds.

Proof. $\bar{x}, \bar{v}$ being the feasible solutions to the constraint sets (6.4), (6.5), it follows that

$$c'\bar{x} = \bar{x}'c \leqslant \bar{x}'A'\bar{v} = (Ax)'\bar{v} = \bar{v}'(Ax) \leqslant \bar{v}'b,$$

or

$$c'\bar{x} \leqslant \bar{v}'b,$$

i.e.

$$f(\bar{x}) \leqslant G(\bar{v}). \tag{6.10}$$

To prove the optimality assertion connected with the relationship (6.9) assume that this is untrue and hence there exists a vector x^* such that this is a solution of A_c and $f(x^*) > f(x^0)$. This implies that $f(x^*) > G(v^0)$, which clearly contradicts the earlier relationship hence the second assertion is proved to be true by "reductio ad absurdum".

(b) The optimal solutions to each of these problems must satisfy the orthogonality or "complementary slackness" property: the definition and proof of this condition follows.

From (6.10) it is observed that for the optimum solution the inequality

$$\bar{x}'c \leqslant \bar{x}'A'\bar{v} \leqslant b'\bar{v},$$

must hold as the equality

$$x^0c = x^0A'v^0 = bv^0. \tag{6.11}$$

From (6.11) it follows $bv^0 - x^0A'v^0 = 0$,

or

$$v^0(b - Ax^0) = 0,$$

i.e.

$$v^0y^0 = 0 \quad \text{or} \quad \sum_{i=1}^{m} v_i^0 y_i^0 = 0. \tag{6.12}$$

Again $\qquad x^0c - x^0A'v^0 = 0$

or $\qquad x^0(c - A'v^0) = 0$

or $\qquad x^0(-u^0) = 0 \quad \text{or} \quad \sum_{j=1}^{n} x_j^0 u_j^0 = 0, \qquad (6.13)$

since, all x, y, u, v are non negative the relations (6.12), (6.13) imply that if a component v_i (u_j) is positive then the corresponding component of $y_i(x_j)$ is zero, and vice versa: this is known as the "Complementary Slackness Theorem". Rewriting $y_i = x_{n+i}; i = 1, 2, \ldots, m$ and $u_{n+i} = v_i; i = 1, 2, \ldots, m$ as indicated earlier (6.12), (6.13) may be combined into one relationship

$$\sum_{j=1}^{m+n} x_j u_j = 0; \qquad x_j, u_j \geqslant 0 \text{ for all } j. \qquad (6.14)$$

(c) In this section cases which exclude the mutual feasibility assumptions of (a) and (b) are discussed.

(i) If the primal problem is feasible and the dual problem is not feasible then the primal problem is not bounded: $f(x) \to +\infty$.
(ii) If the primal problem is not feasible and the dual problem is feasible then the dual has no lower bound: $G(v) \to -\infty$.
(iii) Neither the primal nor the dual problem may have feasible solutions.

TABLE 6.2

	Primal has feasible soln.	Primal has no feasible soln.
Dual has feasible solution	$f(x) \leqslant G(v)$ $\max f(x) = \min G(v)$	$\min G(v) \to -\infty$
Dual has no feasible	$\max f(x) \to +\infty$	This is possible

The possible status of the problem pair are therefore summarized in the Table 6.2.
Some of these properties of the problem are again considered in Section 6.4.

6.3 DUALITY OF THE OTHER FORMS OF LINEAR PROGRAMS

Duality relations for the standard and mixed forms of primal programs may be stated as:

Primal	*Dual*
A_s: Maximize $f(x) = c'x$,	B_s: Minimize $G(v) = b'v$,
subject to $Ax = b$,	subject to $A'v \geqslant C$, (6.15)
and $x \geqslant 0$.	and v unrestricted.
A_m: Maximize $f(x) = \sum_{j=1}^{n} c_j x_j$,	B_m: Minimize $G(v) = \sum_{i=1}^{m} b_i v_i$,
subject to $\sum_{j=1}^{n} a_{ij} x_j \leqslant b_i; i = 1, 2, \ldots, t$,	$\sum_{i=1}^{m} a_{ij} v_i \geqslant c_j; j = 1, 2, \ldots, n$, (6.16)
$\sum_{j=1}^{n} a_{ij} x_j = b_i; i = t+1, \ldots, m$,	$v_i \geqslant 0; i = 1, 2, \ldots, t$,
$x_j \geqslant 0; j = 1, 2, \ldots, n$.	v_i unrestricted in sign for $i = t+1, t+2, \ldots, m$.

The rules for obtaining the dual problem corresponding to the "general form", i.e. the general primal problem, are of most interest and these may be stated as

Primal	*Dual*
Objective form (Max $f(x) = c'x$),	Constant terms: c,
Constant terms: b,	Objective form (Min $G(v) = b'v$),
Coefficient matrix A: a_{ij},	Transposed matrix A': a_{ji},
Constraints:	Variables:
ith inequality: $\sum_j a_{ij} x_j \leqslant b_i$,	ith variable $v_i \geqslant 0$,
ith equality: $\sum_j a_{ij} x_j = b_i$,	ith variable v_i, unrestricted in sign.
Variables:	Constraints:
$x_j \geqslant 0$,	jth inequality: $\sum_i a_{ij} v_i \geqslant c_j$,
x_j unrestricted in sign,	jth equality: $\sum_i a_{ij} v_i = c_j$.

That an equality relation translates into a free dual variable (and vice versa) may be explained in the following simple way. An inequality $\geqslant$ in the primal problem may be retained and the corresponding dual variable is then restricted to non-positive values only. Replacing any primal equality relation by the $\geqslant$ and $\leqslant$ inequalities then introduces two variables in the dual problem non-positive and non-negative respectively, their coefficients in the objective function and the rest of the matrix being the same. This is therefore equivalent to introducing a free variable unrestricted in sign.

6.4 THE DUAL SIMPLEX ALGORITHM

The Dual Simplex Algorithm is more or less symmetric with respect to the primal simplex algorithm and may be stated along with the primal simplex algorithm as follows:

If the initial basic solution is primal (dual) feasible i.e. $\bar{a}_{i0} = \beta_i \geqslant 0,\ i \geqslant 1$ $(\bar{a}_{0j} = \bar{c}_j \geqslant 0; j \geqslant 1)$, choose a column q (row p) with the first element $\bar{a}_{0q}(\bar{a}_{p0})$ negative. From among the positive (negative) elements in this column (row) select the one for which the ratio $\bar{a}_{i0}/\bar{a}_{iq}(\bar{a}_{0j}/\bar{a}_{pj})$ attains its least absolute value. This is the pivot element $\bar{a}_{pq} \neq 0$. If no positive (negative) element exists in such a column (row) the problem is unbounded (has no feasible solution). The optimal solution is obtained when primal feasibility and dual feasibility are attained together.

In the statement of this algorithm there is implicit explanation for the three out of four cases illustrated in Table 6.2.

The primal problem has no feasible solution if one finds an inconsistent row:

$$x_{n+p} = -\beta_p + \sum_{j=1}^{n} \bar{a}_{ij}(-x_j) \quad \text{and} \quad \bar{a}_{ij} \geqslant 0, \tag{6.17}$$

for such an equation represented by this row no positive value of x_j can drive the infeasible value $-\beta_p$ of the variable x_{n+p} positive hence feasible. The dual simplex row choice rule discovers this situation similarly, the dual problem has no feasible solution and the primal is unbounded if all the entries in a column q with negative reduced cost is such that $\bar{a}_{iq} \leqslant 0$ for $i \geqslant 1$.

If in a tableau both these conditions hold see (Table 6.3) whereby the

TABLE 6.3

		$-\bar{c}_q$				
		$-$				
$-\beta_p$	$+$	$+$	0	$+$	$+$	$+$
		$-$				
		$-$				

common element at the intersection of the row and column is zero. Then both the primal and the dual are not feasible and the problem is not meaningful in either form. This therefore illustrates the fourth case in Table 6.2.

6.5 AN EXAMPLE ILLUSTRATING THE DUALITY PROPERTIES

In this section some of the properties of the primal and dual problem pair are illustrated by means of an example.

Consider the problem (primal problem),

$$
\begin{aligned}
&\text{Maximize } f(x) = 4x_1 + 3x_2 \\
&\text{subject to} \quad 6x_1 + 2x_2 \leqslant 12 \\
&\qquad\qquad\quad 2x_1 + 2x_2 \leqslant 5 \\
&\qquad\qquad\qquad\quad x_2 \leqslant 2 \\
&\text{and} \qquad\quad x_1, x_2 \geqslant 0.
\end{aligned} \tag{6.18}
$$

The dual to the problem can be stated as

$$
\begin{aligned}
&\text{Minimize } G(u) = 12u_3 + 5u_4 + 2u_5 \\
&\text{subject to} \quad 6u_3 + 2u_4 \geqslant 4 \\
&\qquad\qquad\quad 2u_3 + 2u_4 + u_5 \geqslant 3 \\
&\text{and} \qquad\quad u_3, u_4, u_5 \geqslant 0.
\end{aligned} \tag{6.19}
$$

These two problems are set up in the Tableau 6.4 in which x_3, x_4, x_5 are the slack variables for the primal problem and u_1, u_2 are the surplus variables for the dual problem. Since this is a primal feasible tableau one normally applies the primal simplex algorithm and through the sequence of Tableaux 6.5, 6.6 the optimum solution is obtained. However, if the pivotal exchange $x_1|x_4$ is carried out in Tableau 6.4 and $x_2|x_5$ in Tableau 6.5 then the neighbouring Tableaux 6.7 and 6.8 are obtained. Both these tableaux are dual feasible. If the dual simplex algorithm is applied to both these tableaux then the optimum Tableau 6.6 is obtained from both these tableaux in one step.

The Table 6.9 summarises some pertinent features of these tableaux and the Fig. 6.1 provides a graphical representation of the state corresponding to each one of these tableaux. Of the five tableaux, two are primal feasible, two are dual feasible and one tableau is both primal and dual feasible, i.e. optimum.

	$G(u)$	$-x_1$	$-x_2$	
$f(x)$	0	-4	-3	
x_3	12	6	2	u_3
x_4	5	2	2	u_4
x_5	2	0	1	u_5
		u_1	u_2	

TABLEAU 6.4

$\xrightarrow{x_1/x_4}$ (to Tableau 6.7); $x_1|x_3$ (down to Tableau 6.5)

	$G(u)$	$-x_4$	$-x_2$	
$f(x)$	10	2	1	
x_3	-3	-3	-4	u_3
x_1	5/2	1/2	1	u_1
x_5	2	0	1	u_5
		u_4	u_2	

TABLEAU 6.7

	$G(u)$	$-x_3$	$-x_2$	
$f(x)$	8	2/3	$-5/3$	
x_1	2	1/6	1/3	u_1
x_4	1	$-1/3$	4/3	u_4
x_5	2	0	1	u_5
		u_3	u_2	

TABLEAU 6.5

$\xrightarrow{x_2|x_5}$ (to Tableau 6.8); $x_2|x_4$ (down to Tableau 6.6)

	$G(u)$	$-x_3$	$-x_5$	
$f(x)$	34/3	2/3	5/3	
x_1	4/3	1/6	$-1/3$	u_1
x_4	$-5/3$	$-1/3$	$-4/3$	u_4
x_2	2	0	1	u_2
		u_3	u_5	

TABLEAU 6.8

Dual simplex $x_2\backslash x_3$ (Tableau 6.7 to Tableau 6.6); Dual simplex $x_5\backslash x_4$ (Tableau 6.8 to Tableau 6.6)

	$G(u)$	$-x_3$	$-x_4$	
$f(x)$	37/4	1/4	5/4	
x_1	7/4	1/4	$-1/4$	u_1
x_2	3/4	$-1/4$	3/4	u_2
x_5	5/4	1/4	$-3/4$	u_5
		u_3	u_4	

TABLEAU 6.6

Observe the following inequalities corresponding to these tableaux

$$(0, 8) \leqslant 37/4 \leqslant (10, 34/3),$$

$$\begin{matrix}\text{Primal} \\ \text{feasible}\end{matrix} \leqslant \left\{\begin{matrix}\text{Primal and} \\ \text{dual} \\ \text{feasible}\end{matrix}\right\} \leqslant \begin{matrix}\text{Dual} \\ \text{feasible}\end{matrix} \qquad (6.20)$$

Optimum

The relationship of (6.20) translates directly in numerical terms the inequalities derived in section 6.2.

Finally, a pertinent geometric interpretation associated with the dual simplex algorithm is considered here. It assumes importance in the cutting plane theory of integer programming considered in chapter 7.

TABLE 6.9.

Tableau	Feasible		Objective	Solution values									
	Primal?	Dual?	$f(x)/G(u)$	x_1	x_2	x_3	x_4	x_5	u_1	u_2	u_3	u_4	u_5
6.4	Yes	No	0	0	0	12	5	2	−4	3	0	0	0
6.5	Yes	No	8	2	0	0	1	2	0	−5/3	2/3	0	0
†6.6	Yes	Yes	37/4	7/4	3/4	0	0	5/4	0	0	1/4	5/4	0
6.7	No	Yes	10	5/2	0	−3	0	2	0	1	0	2	0
6.8	No	Yes	34/3	4/3	2	0	−5/3	0	0	0	2/3	0	5/3
.													
.													
.													

† Optimal solution.

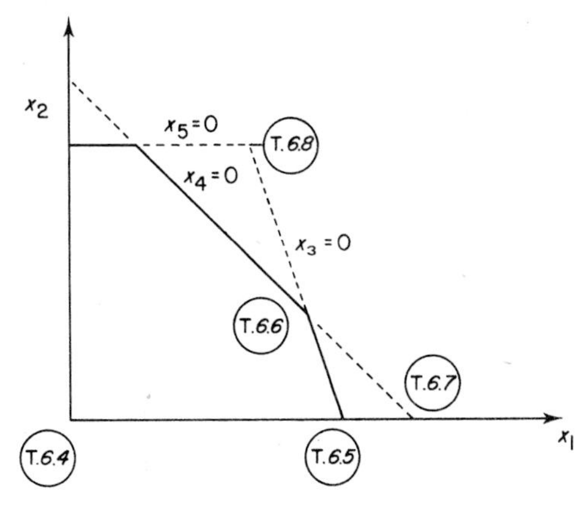

FIG. 6.1

If in the original problem the second constraint

$$2x_1 + 2x_2 \leqslant 5,$$

were ignored then the Tableau 6.8 would have contained the optimal solution to the problem. However, taking into consideration this constraint the solution is not feasible and hence not the optimum solution in as much as that the plane (line) $x_4 = 0$ separates it from the feasible convex region. A dual simplex step $x_5|x_4$ therefore may be given the geometric interpretation of making such a "cutting plane" active and resolving the infeasibility associated with it.

6.6 ECONOMIC INTERPRETATION OF DUALITY

The standard linear programming problem admits interpretation in very broad terms which fit into a number of models of applied economics. In general a maximization problem represents the goal of an enterpreneur who wishes to maximize his profit; the minimization problem on the other hand represents the objective of a consumer or a central authority to minimize cost.

For a primal maximization problem

$$\begin{aligned} &\text{Maximize } c'x \\ &\text{subject to } Ax \leqslant b, \\ &\qquad\qquad x \geqslant 0, \end{aligned} \tag{6.21}$$

the non-negative value of the jth activity i.e. $x_j \geqslant 0$ represents the level at which the product j must be produced. a_{ij} the element of the technology matrix represents the quantity of commodity i required in the product j at unit level, b_i represents the availability of the commodity i and c_j represents the profit corresponding to unit of product j. The problem pair associated with these coefficients is set out below.

Primal problem

$\sum_{j=1}^{n}$ (level of product j) $\times$ (production rate of j at unit level from commodity i)

$\leqslant$ (availability i), $\quad i = 1, 2, \ldots, m$

Level of product $j \geqslant 0$

Maximize (overall profit) $= \sum_{j=1}^{n}$ (price per unit of product j) $\times$ (level of j).

Dual problem

$\sum_{i=1}^{m}$ (Unit cost of commodity i) $\times$ (production rate of j at unit level from commodity i) $\geqslant$ (price per unit of product j)

Unit cost of commodity $i \geqslant 0$

Minimize (cost) $= \sum_{i}$ (availability of commodity i) $\times$ (unit cost of i)

Thus these two problems represent two opposing interests. The former addresses itself to the question of determining the level of production (x_j?) the latter of fixing suitable prices (u_i?). Note the solution of one naturally solves the other problem.

6.7 POST OPTIMAL ANALYSIS: RANGING AND PARAMETRIC PROCEDURES

When linear programming is being used as a serious modelling tool the interest in the model does not stop just as the optimal feasible solution is obtained. Having got to the optimum solution the most searching analysis of the model starts and naturally the questions arise as to how will the solutions be affected if the coefficients defining the system were changed. Investigations which may be carried out for such post optimal analysis have been defined in a systematic fashion.

The post optimal analysis falls in two main classes. If only one component of the coefficients defining the problem is altered at a time then this is called "Ranging Procedure" or "Sensitivity Analysis". If a second vector of right-hand side or a cost row is used and the optimum solution to a parametric variation of composite cost and rhs are investigated then this is known as "Parametric Programming".

Ranging procedures

Depending on the nature of the coefficients which are analyzed ranging may be divided into three sub-classes, viz., cost ranging, r.h.s. ranging, A-ranging. The last stands for ranging the coefficients of the A-matrix and is not considered here.

Consider the problem

$$\text{Maximize } c'x$$
$$\text{subject to } Ax \leqslant b,$$
$$x \geqslant 0.$$

Cost ranging. If one coefficient at a time of the objective function is taken and the sensitivity of the optimal solution to the possible range of values of this coefficient is investigated then this is known as cost ranging. Two different situations need to be considered.

Let the variable x_p corresponding to the coefficient c_p of the objective function be nonbasic. Let the reduced cost coefficient of this variable in the optimal tableau be $\bar{c}_p$; this implies that $\bar{c}_p \geqslant 0$. Then it is obvious that the upper critical value c_p is $c_p + \bar{c}_p$ beyond which x_p requires to be brought into the basis. There is of course no lower limit to c_p in this case.

On the other hand if this variable in the optimal solution of the definitive problem is a basic variable then the upper and lower critical values of the range of the cost coefficient c_p may be computed in the following way. Let $\bar{c}_{j_1}, \bar{c}_{j_2}, \ldots, \bar{c}_{j_n}$ denote the reduced cost coefficients of the nonbasic variables $x_{j_1}, \ldots, x_{j_n}$, in the optimal tableau. Let $\bar{a}^p_{ij_1}, \bar{a}^p_{ij_2}, \ldots, \bar{a}^p_{ij_n}$ denote the tableau coefficients of the ith row in which the variable x_p is pivoted. Let PS denote the set of indices of positive elements i.e., $PS = \{j_k | \bar{a}^p_{ij_k} > 0\}$ similarly let NG be defined as $NG = \{j_k | \bar{a}^p_{ij_k} < 0\}$. Let the dual pivot choice rule be applied to this row and the minimum ratio be obtained for the column index j_q such that

$$\bar{c}_{j_q} / |\bar{a}^p_{ij_q}| = \min_{j_k \in NG} \bar{c}_{j_k} / |\bar{a}^p_{ij_k}|. \tag{6.22}$$

Pivoting on the element $\bar{a}^p_{ij_q}$ provides a dual feasible tableau. In this tableau $\bar{c}_p = -\bar{c}_{j_q} / \bar{a}^p_{ij_k}$ which is a nonnegative quantity.

Consider the pivotal step back from this tableau into the optimal tableau. It may be easily shown that if c_p is changed to $c_p - \bar{c}_{j_q} / \bar{a}^p_{ij_q}$ then this provides a critical value above which the present tableau is no longer optimal.

Similarly the lower critical value may be deduced as

$$c_p - \bar{c}_{j_r}/\bar{a}^p_{ij_r} \quad \text{where} \quad \bar{c}_{j_r}/\bar{a}^p_{ij_j} = \min_{j_k \in PS} \bar{c}_{j_k}/\bar{a}^p_{ij_k}. \tag{6.23}$$

It is apparent from the column choice rules of (6.22) and (6.23) that these critical values may be obtained by applying the analysis of one dual step of the simplex algorithm.

R.H.S. ranging.

The sensitivity of the optimal feasible solution to the variation of any component of the r.h.s. vector is considered for two different cases.

Assume that the slack vector x_{n+p} associated with the inequality p is in the optimal basis pivoted in the row i, and has a value β^p_i and further let the initial r.h.s. be b_p. Then the lower critical value of the component is $b_p - \beta^p_i$ and it has no upper critical value. However, if the slack x_{n+p} is nonbasic in the optimal tableau then there may be two critical values for the corresponding pth component b_p of the r.h.s. vector. Let $\bar{a}_{ij_p}$, $i = 1, 2, \ldots, m$ be the tableau entries corresponding to this slack column. Again define two sets of row indices $PS = \{i | \bar{a}_{ij_p} > 0\}$ and $NG = \{i | \bar{a}_{ij_p} < 0\}$ for the elements in this column. Then the critical values are given by the expressions

$$\text{Lower critical value} = b_p - \beta_r/\bar{a}_{rj_p} \tag{6.24}$$

$$\text{Upper critical value} = b - \beta_s/\bar{a}_{sj_p} \tag{6.25}$$

where

$$\beta_r/\bar{a}_{rj_p} = \min_{i \in PS} \beta_i/\bar{a}_{ij_p} \tag{6.26}$$

and

$$\beta_s/|\bar{a}_{ij_p}| = \min_{i \in NG} \beta_i/|\bar{a}_{ij_p}| \tag{6.27}$$

The relations (6.24)...(6.27) are deduced by applying the analysis of one primal step of the simplex algorithm.

Parametric procedures

Parametric analysis may be subdivided into three different subclasses viz., Paracost, Pararhs, Pararim. These are defined in this section; the algorithms for carrying out these analyses are discussed in [2] and [43].

Consider the problem

$$\text{Maximize } c'x \tag{6.28}$$

$$\text{subject to } Ax \leqslant b \tag{6.29}$$

$$x \geqslant 0,$$

and two further vectors $(c^*)'$; $1 \times n$ for a second cost row and b^*, $m \times 1$ for a second r.h.s. The definitions that follow also hold for the problems stated in other forms.

Paracost. The "parametric cost" investigation of this model consists of looking at the discrete steps of the parameter θ and the corresponding solution values, (primal and dual) of the problem maximize $f(x) = (c + \theta c^*)'x$ subject to the constraints (6.29). The investigation starts at the optimal solution for (6.28), (6.29) and $\theta = 0$ and may terminate:

(a) if the parameter θ becomes unbounded i.e. for further increase in the value θ the current optimal basis continues to be optimal,
(b) if for the current value of θ the corresponding solution becomes unbounded.

Pararhs. The "parametric r.h.s." investigation of the model consists of looking at the discrete steps of θ and the corresponding solution values of the problem maximize $f(x) = c'x$ subject to $Ax \leqslant (b + \theta b^*)$ and $x \geqslant 0$.

The investigation starts at the optimal value of the definitive problem and terminates

(a) if the parameter θ becomes unbounded i.e. for further increase in the value of θ the current basis continues to be optimal,
(b) if for the current value of θ and beyond the constraint set becomes non feasible.

Pararim. This procedure is also known as the "composite parametrics" and consists of investigating the problem

$$\begin{aligned} &\text{Maximize } f(x) = (c + \theta c^*)'x \\ &\text{subject to} \qquad Ax \leqslant (b + \theta b^*) \\ &\text{and} \qquad x \geqslant 0. \end{aligned} \tag{6.30}$$

This is clearly a combination of the previous two analyses in one and may terminate in one of the above three ways.

EXERCISES

6.1 State the dual of the problem,

$$\begin{aligned} \text{Maximize} \quad & f(x) = 2x_1 + 3x_2 + 2x_3 \\ \text{subject to} \quad & 2x_1 + 5x_2 + x_3 \leqslant 25 \\ & x_1 + x_2 - x_3 = 15 \\ & x_1, x_2 \geqslant 0 \text{ and } -\infty < x_3 < +\infty. \end{aligned}$$

6.2 What is the most appropriate algorithm to solve the following problem?

Minimize $z = 2x_1 + x_2 + x_3$
subject to

$$\begin{aligned} 4x_1 + 6x_2 + 3x_3 + x_4 &= 8 \\ x_1 - 9x_2 + x_3 + x_5 &= -3 \\ -2x_1 - 3x_2 + 5x_3 + x_6 &= -3 \end{aligned}$$

and $x_1, x_2, x_3, x_4, x_5, x_6 \geqslant 0$.

If the objective function of the above problem is changed to

minimize $z = 2x_1 + x_2 - x_3$

and one still wishes to apply the previous algorithm, then suggest a method of appropriately transforming the problem. Solve both these problems.

6.3 Devise a procedure (by extending the dual simplex algorithm) for tracing the optimal solutions of the problem,

$$\text{Maximize} \quad x_0 = \sum_{j=1}^{n} c_j x_j$$

$$\text{subject to} \quad \sum_{j=1}^{n} a_{ij} x_j = b_i + \theta b_i^*, \quad (i = 1, 2, \ldots, m), \quad 0 \leqslant \theta \leqslant \theta_{\max}$$

and $x_j \geqslant 0, j = 1, 2, \ldots, n$.

Describe the procedures r.h.s. ranging, cost ranging and A ranging and some contexts in which these are used.

6.4 At a certain stage of solving a linear programming problem, the contracted tableau appears as shown below.

		$-x_1$	$-x_2$	$-x_4$
x_0	12	1	-5	2
x_3	6	3	-3	5
x_5	-5	6	0	6
x_6	2	5	-6	5

What can you deduce above the feasibility and/or the boundedness of this problem and its dual?

6.5 Devise a dual simplex algorithm which uses the product form of inverse of the basis matrix. If implicitly upper bounded variables are used state how this algorithm should be modified.

CHAPTER 7

Integer and Mixed Integer Linear Programs

KEYWORDS: Pure Integer Problem, All Integer Integer Programming Problem, Mixed Integer Problem, Cutting Planes, Integer Forms, Dichotomy, Branching Variable, Subproblem, Penalties, Tree Search.

7.1 STATEMENT AND CLASSIFICATION OF PROBLEMS

The integer programming problem may be simply stated as a linear program in which some or all the variables must be integer. When all the variables are constrained to be integer, the linear objective function and the linear constraint functions are defined over an integer domain (the lattice points in the continuous linear space). When only some variables are constrained to be integer there exists a continuous space over which the rest of the variables are defined. In either case, however, the corresponding problem is not linear: a simple test being that all the partial derivatives of each of these functions do not exist (in a linear problem the partial derivatives are defined for all the component directions and are constant for each continuous linear function).

The terminology used in identifying integer programming problem types is not rigorous; the classification made in this section is not necessarily the most standard; however, it serves to define most of the terminology in vogue.

If to the statement of the linear programming problem:

(i)

$$\text{maximize } f(x) = \sum_{j=1}^{n} c_j x_j \tag{7.1}$$

$$\text{subject to} \quad \sum_{j=1}^{n} a_{ij} x_j \leqslant b_i \tag{7.2}$$

$$\text{and} \quad x_j \geqslant 0, \quad j = 1, 2, \ldots, n; \tag{7.3}$$

the further constraint,

$$x_j \text{ integer for } j \in J, \text{ where } J = \{1, 2, \ldots, n\} \tag{7.4}$$

is added, then this is called the (pure) integer programming problem. If m slack variables x_{n+i}, $i = 1, 2, \ldots, m$ are introduced and to (7.1), (7.2) n-trivial equations $x_j = 0 - 1(-x_j)$; $j = 1, 2, \ldots, n$, are added then the system may be rewritten in the Tucker–Beale form:

(ii) $$\text{maximize } f(x) = 0 + \sum_{j=1}^{n} - c_j(-x_j)$$

$$\text{subject to } x_{n+i} = b_i + \sum_{j=1}^{n} a_{ij}(-x_j), \quad i = 1, 2, \ldots, m \tag{7.5}$$

$$x_j = 0 + (-1)(-x_j), \qquad j = 1, 2, \ldots, n$$

$$x_j \geqslant 0, \qquad j = 1, 2, \ldots, n + m$$

and $$x_j \equiv 0 \bmod (1)\dagger, \qquad j \in J, \tag{7.6}$$

where J as in (7.4), denotes the set of first n indices identifying the integer variables. This is also the most popular tableau used for hand computation of the integer programming problems.

(a) *Pure integer problem*

The statement (i) of the integer programming problem where all the variables with indices in the set J are constrained to take integer values, specifies a Pure Integer Problem.

(b) *All integer integer problem*

This is a special case of Pure Integer Problem hence the statement (ii) of the problem holds and further,

$$\begin{aligned} a_{ij} &\equiv 0 \bmod (1) \qquad \text{for all } i, j \\ c_j &\equiv 0 \bmod (1) \qquad \text{for all } j \\ b_i &\equiv 0 \bmod (1) \qquad \text{for all } i. \end{aligned} \tag{7.7}$$

This problem has the interesting connotation that by definition all the stack variables x_{n+i} in the problem are integer (and of course non negative). Any pure integer problem which has the coefficients of each row of the matrix made up of rational numbers may be transformed to an all integer integer

† If a, b, c are integers then $a \equiv b \bmod (c)$ means that $(a\text{–}b)$ is divisible by c; further $a \equiv 0 \bmod 1$ implies that a is an integer.

problem by multiplying each row by the least common multiplier of the denominators. The corresponding (statement (ii)) problem is then defined in non-negative integer slack variables, and original integer variables.

(c) 0–1 *Integer problem*

An all integer integer programming problem where all the variables structural or slack can take only 0–1 value is said to be a 0–1 integer programming problem.

(d) *Mixed integer problem*

If in the statement (ii) for an all integer integer programming problem the constraint (7.6) is replaced by the constraint

$$x_j \equiv 0 \bmod (1), \qquad j \in K \tag{7.8}$$

where $K \subset J$ (K is a proper subset of J) then the problem is called a mixed integer problem.

It is interesting to note that the pure integer programming problem which is not an all integer integer programming problem, when stated in form (ii) is in some sense a mixed integer problem since the slack variables are not necessarily integer for any feasible solutions of the problem (see Section 7.4).

One may observe that any problem (c) is also (b) and any problem (b) is also a problem (a); if K in (7.8) is defined as a subset of J (rather than a proper subset of J) then any problem (a) is also a problem (d). This implies that any algorithm that solves (d) will also solve (a), (b), (c), however, any algorithm that solves (c) may not solve (b) etc.

7.2 CLASSIFICATION OF ALGORITHMS

A variety of algorithms have been developed to solve different types of integer programming problems [31] and these may be classified as shown in Table 7.1. Of the different algorithms listed in this table only the Dual Cutting Plane Method and the Branch and Bound Method are considered here.

7.3 DUAL CUTTING PLANE METHOD OR THE METHOD OF INTEGER FORMS

This method due to Gomory [24] was first published in 1958 and marks the turning point in the theory of the algorithms for solving integer programming

TABLE 7.1

Problem types	Algorithms for their solution
1. (d), (a)	1. Primal Cutting Plane Methods. ⟨Limited computational results have been reported⟩
2. (d), (a)	2. Dual Cutting Plane Methods ⟨Some computational results have been reported; not always encouraging⟩
3. (c)	3. Implicit Enumeration Method ⟨Some computational results have been reported; not always encouraging⟩
4. (b)	4. All Integer Integer Programming Algorithm (a variation of (2)) ⟨Limited computational results have been reported⟩
5. (d)	5. Partitioning Methods ⟨Some computational results have been reported and have proven to be of limited success⟩
6. (d), (a)	6. Branch and Bound Methods ⟨Extensive computational results have been reported. It is the most accepted method for solving such problems⟩

problems. Only a concise statement of the algorithm as applied to pure integer problems is provided in this section. For convenience and also to illustrate some properties of integer programs, the all integer case of the pure integer problem is considered. For a proper theoretical discussion of the method and convergence proof etc. see [15], [24], [27], [45].

Let $C(A)$ denote the solution to the associated continuous linear problem stated in (7.5) and let $I(A)$ denote the optimal integer solution to the integer problem stated in (7.5), (7.6). If the all integer property (7.7) is further assumed then it is possible to obtain an integer, D which is the absolute value of the determinant of the basis of the continuous optimum solution. The continuous optimum solution may be expressed in the form,

$$x_0 = \bar{a}_{00} + \sum_{q_j \in N} \bar{a}_{0j}(-x_{q_j}) \tag{7.9}$$

$$x_i = \bar{a}_{i0} + \sum_{q_j \in N} \bar{a}_{ij}(-x_{q_j}) \qquad i = 1, 2, \ldots, n + m,$$

or in the form,

$$x_0 = \alpha_{00}/D + \sum_{q_j \in N} \alpha_{i0}(-x_{q_j})/D$$

$$x_i = \alpha_{i0}/D + \sum_{q_j \in N} \alpha_{ij}(-x_{q_j})/D, \qquad i = 1, 2, \ldots, n + m, \tag{7.10}$$

where N is the set of indices of the non-basic variables and $\alpha_{ij} \equiv 0 \bmod (1)$

are integers for all i, j. The statements (7.9), (7.10) imply that every coefficient $\bar{a}_{ij}$ of the tableau for the continuous optimum solution, may be expressed as a ratio of two integers α_{ij}/D, the common denominator in each case is D. If $\alpha_{i0} \not\equiv 0 \bmod (D)$ for any $i = 1, 2, \ldots, m + n$ then $C(A) \not\equiv I(A)$ since x_i has the value $\alpha_{i0}/D = \bar{a}_{i0}$, etc. Also observe the bound $C(A) \geqslant I(A)$.

Prior to describing the inequality derived by Gomory, a proposal due to Dantzig is considered. If all the x_{q_j} ($q_j \in N$) are set to their non-basic zero values then the continuous optimum solution values α_{i0}/D do not satisfy the integer constraint for all i. At least one of the variables x_{q_j} therefore must assume non-zero value in a feasible integer solution to the problem hence the inequality (Dantzig-cut),

$$\sum_{q_j \in N} x_{q_j} \geqslant 1 \tag{7.11}$$

must hold, this is because the variables x_{q_j} are themselves non-negative and must be integral. By adding such additional rows to the problem and solving the problem again the possible integer solution to the original problem may be investigated. However, this procedure may not converge and the method due to Gomory, the description of which follows has proven to be more sound and successful.

In the continuous optimum solution (7.9), (7.10) for a row corresponding to an integer-infeasible variable, x_e, $\bar{a}_{e0} \not\equiv 0 \bmod (1)$, or $\alpha_{e0} \not\equiv 0 \bmod (D)$ and with at least another element $\bar{a}_{ej} \not\equiv 0 \bmod (1)$, or $\alpha_{ej} \not\equiv 0 \bmod (D)$, define the fractional elements;† and

$$\left.\begin{aligned} f_{ej} &= \bar{a}_{ej} - [\bar{a}_{ej}] \\ f_{e0} &= \bar{a}_{e0} - [\bar{a}_{e0}] \end{aligned}\right\} \quad q_j \in N. \tag{7.12}$$

Note that these may also be expressed as ratio of integers,

$$f_{ej} = \frac{d_{ej}}{D} \quad \text{and} \quad f_{e0} = \frac{d_{e0}}{D}, \tag{7.13}$$

where

$$d_{ej} = D \times \left\{\frac{\alpha_{ej}}{D} - \left[\frac{\alpha_{ej}}{D}\right]\right\} \quad \text{and} \quad d_{e0} = D \times \left\{\frac{\alpha_{e0}}{D} - \left[\frac{\alpha_{e0}}{D}\right]\right\}. \tag{7.14}$$

† If p is a real number then the quantity $[p]$ is defined as the mathematical integral part of p, such that $[p]$ is the largest integer less than or equal to p. This also defines a positive fraction associated with this real quantity p such that

$$p = [p] + f \quad \text{where} \quad \leq f < 1.$$

Thus for $\quad p = 1{\cdot}3;\ [p] = 1{\cdot}0, f = 0{\cdot}3,$

and for $\quad p = -1{\cdot}3;\ [p] = -2{\cdot}0, f = 0{\cdot}7.$

The variable x_e (in the row e of the equation) expressed in terms of the non-basic variables may be written in the form,

$$x_e = [\bar{a}_{e0}] + f_{e0} + \sum_{q_j \in N} [\bar{a}_{ej}](-x_{q_j}) + \sum_{q_j \in N} f_{ej}(-x_{q_j}). \qquad (7.15)$$

In the right-hand side of (7.15) the expression for x_e may be presented in two parts;

(i) $$[\bar{a}_{e0}] + \sum_{q_j \in N} [\bar{a}_{ej}](-x_{q_j}), \text{ and} \qquad (7.16)$$

(ii) $$f_{e0} + \sum_{q_j \in N} f_{ej}(-x_{q_j}). \qquad (7.17)$$

The expression (7.16) in (i) of course leads to an integer quantity, since all the components which are summed are themselves products of integer quantities; however, the resulting integers may not necessarily be non-negative. For any non-integer solution value of x_e the expression (7.17) in (ii) will assume a value which must be sum of a non-positive quantity and $\sum_{q_j \in N} f_{ej}(-x_{q_j})$ and a positive fraction f_{e0}.

For an integer value of x_e therefore the quantity

$$f_{e0} + \sum_{q_j \in N} f_{ej}(-x_{q_j})$$

must take integer value and this may be expressed in terms of the relationship

$$\sum_{q_j \in N} f_{ej}(x_{q_j}) \geqslant f_{e0} \qquad (7.18)$$

or

$$\sum_{q_j \in N} f_{ej}(x_{q_j}) - f_{e0} - s_e = 0$$

$$s_e = -f_{e0} + \sum_{q_j \in N} f_{ej}(x_{q_j}) \qquad (7.19)$$

where s_e must be non-negative and integer.

The equation derived in (7.19) must be satisfied by any feasible integer solution to the original problem. It also has the effect of excluding the current optimum solution to the associated continuous problem. The inequality (7.18) or the equation (7.19) are said to represent a cutting plane. In Gomory's method a series of such cutting planes are introduced and the series of associated continuous problems are solved:

$$C^0(A) \geqslant C^1(A) \geqslant C^2(A) \ldots \geqslant I(A)$$

until the integer requirement is satisfied and the corresponding integer solution $I(A)$ is obtained.

The algorithm of integer form may be stated as

(i) Solve the associated continuous problem $C^p(A)$ [$p = 0$] by an appropriate primal or dual algorithm. If "no feasible" or "unbounded" condition is met, go to *Exit*.
(ii) If the continuous optimum solution satisfies integer requirement then $C^p(A) = I(A)$; [p may be $0, 1, \ldots, p$ being a count of the cuts and the associated problems generated] in this case go to *Exit*.
(iii) For any $x_e = \bar{a}_{e0}$, $\bar{a}_{e0} \not\equiv 0 \bmod (1)$, and at least one other row element $\bar{a}_{ej} \not\equiv 0 \bmod (1)$, $q_j \in N$, derive the cutting plane,

$$s_e = f_{e0} - \sum_{q_j \in N} f_{ej}(-x_{q_j}) \text{ as in (7.19)}$$

which introduces a primal infeasibility to an otherwise dual feasible (optimal) tableau set $p = p + 1$.
(iv) Apply dual simplex algorithm to solve the resulting problem. If an optimal solution is obtained go to step (ii), otherwise a no feasible condition is detected go to *Exit*.
(v) *Exit*. The initial problem may be non-feasible or unbounded. If control is transferred from step (ii) an optimum integer solution is obtained. Otherwise the problem has no integer feasible solution.

7.4 AN EXAMPLE ILLUSTRATING THE METHOD OF INTEGER FORMS

Consider the problem,

$$\begin{aligned} &\text{Maximize } 3x_1 + 4x_2 \\ &\text{subject to } \tfrac{2}{5}x_1 + x_2 \leqslant 3 \\ &\tfrac{2}{5}x_1 - \tfrac{2}{5}x_2 \leqslant 1 \qquad (7.20) \\ &x_1, x_2 \geqslant 0 \quad \text{and} \quad x_1, x_2 \text{ integer.} \end{aligned}$$

Although this problem may be loosely referred to as a pure integer problem, if one introduces the slack variables the corresponding problem;

$$\begin{aligned} &\text{Maximize } 3x_1 + 4x_2 \\ &\text{subject to } \tfrac{2}{5}x_1 + x_2 + x_3 = 3 \\ &\tfrac{2}{5}x_1 - \tfrac{2}{5}x_2 + x_4 = 1, \qquad (7.21) \\ &x_1, x_2, x_3, x_4 \geqslant 0 \text{ and } x_1, x_2 \text{ integer,} \end{aligned}$$

is a mixed integer problem since integer values of x_1, x_2 do not imply that

x_3, x_4 must be integer. However, if the two inequalities in (7.20) are multiplied by the common denominator 5 then this leads to the all-integer problem,

$$\begin{aligned} &\text{Maximize } 3x_1 + 4x_2 \\ &\text{subject to } 2x_1 + 5x_2 \leqslant 15 \\ &\qquad\qquad 2x_1 - 2x_2 \leqslant 5 \\ &x_1, x_2 \geqslant 0 \text{ and } x_1, x_2 \text{ integer.} \end{aligned} \tag{7.22}$$

If slack variables x_3, x_4 are introduced, the corresponding problem,

$$\begin{aligned} &\text{Maximize } 3x_1 + 4x_2 \\ &\qquad 2x_1 + 5x_2 + x_3 = 15 \\ &\qquad 2x_1 - 2x_2 + x_4 = 5 \\ &x_1, x_2, x_3, x_4 \geqslant 0, x_1, x_2 \text{ integer} \end{aligned} \tag{7.23}$$

is also a pure integer problem since integer values of x_1, x_2 imply x_3, x_4 must also take on integer values.

The transformed problem as expressed in the form (7.23) is set up in the Tucker–Beale Tableau: 7.1. Applying the primal simplex algorithm the continuous optimum solution is obtained as displayed in Tableau 7.3. The slacks s_1, s_2, s_3 represent the cutting planes introduced in the Tableaux 7.4, 7.5, 7.6, lead to primal infeasibility, and exclude the corresponding optimal solutions. The Tableau 7.7. contains the optimal integer solution. The slack variables s_1, s_2, s_3 are expressed in terms of the structural variables x_1, x_2 (next to the tableau) and the corresponding cutting planes are illustrated in Fig. 7.1.

		$-x_1$	$-x_2$
x_0	0	-3	-4
x_3	15	2	(5)
x_4	5	2	-2
x_1	0	-1	0
x_2	0	0	-1

TABLEAU 7.1

		$-x_1$	$-x_3$
x_0	12	$-7/5$	4/5
x_3	0	0	-1
x_4	11	(14/5)	2/5
x_1	0	-1	0
x_2	3	2/5	1/5

TABLEAU 7.2

		$-x_4$	$-x_3$
x_0	35/2	1/2	1
x_3	0	0	-1
x_4	0	-1	0
x_1	55/14	5/14	1/7
x_2	10/7	$-1/7$	1/7

TABLEAU 7.3

		$-x_4$	$-x_3$
x_0	35/2	1/2	1
x_3	0	0	-1
x_4	0	-1	0
x_1	55/14	5/14	1/7
x_2	10/7	$-1/7$	1/7
s_1	$-13/14$	$(-5/14)$	$-1/7$

TABLEAU 7.4

Cutting plane from the row of x_1 variables

$$5/14x_4 + \frac{x_3}{7} \geqslant 13/14$$

$$s_1 = -13/14 + 5/14x_4 + \frac{x_3}{7}$$

$$= -13/14 + \tfrac{5}{14}(5 - 2x_1 + 2x_2) + \tfrac{1}{7}(15 - 2x_1 - 5x_2)$$

$$= 3 - x_1; x_1 + S_1 = 3 \text{ or } x_1 \leqslant 3$$

		$-s_1$	$-x_3$
x_0	81/5	7/5	4/5
x_3	0	0	-1
x_4	13/5	$-14/5$	2/5
x_1	3	1	0
x_2	9/5	$-2/5$	1/5
s_2	$-4/5$	$(-3/5)$	$-1/5$

TABLEAU 7.5

Cutting plane from the row of x_2 variables

$$\tfrac{3}{5}s_1 + \frac{x_3}{5} \leqslant 4/5$$

$$s_2 = -4/5 + 3/5\, s_1 + x_3/5$$

$$= -4/5 + 3/5\,(3 - x_1) + \tfrac{1}{5}\,(15 - 2x_1 - 5x_2)$$

$$= 4 - x_1 - x_2; \text{ i.e., } x_1 + x_2 \leqslant 4$$

		$-s_2$	$-x_3$
x_0	43/3	7/3	1/3
x_3	0	0	-1
x_4	19/3	$-14/3$	4/3
x_1	5/3	5/3	$-1/3$
x_2	7/3	$-2/3$	1/3 ←
s_3	$-1/3$	$-1/3$	$(-1/3)$

TABLEAU 7.6

Cutting plane from the row of x_2 *variable*

$$\tfrac{1}{3}s_2 + \tfrac{1}{3}x_3 \geqslant 1/3; \qquad s_3 = -1/3 + s_2/3 + x_3/3$$

$$s_3 = 6 - x_1 - 2x_2; \text{ i.e., } x_1 + 2x_2 \leqslant 6$$

		$-s_2$	$-s_3$
x_0	14	2	1
x_3	1	1	-3
x_4	5	-6	4
x_1	2	2	-1
x_2	2	-1	1

TABLEAU 7.7

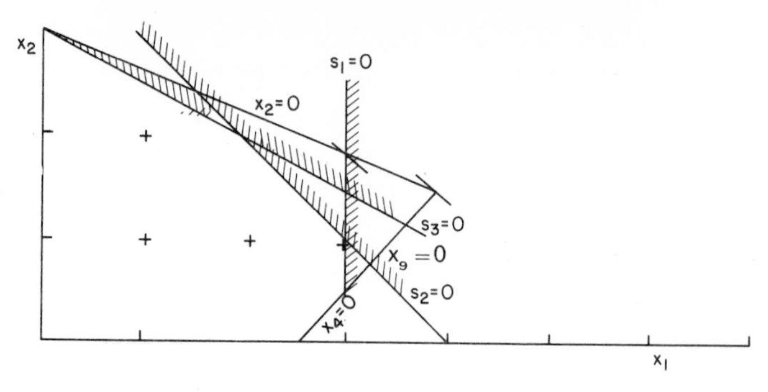

FIG. 7.1

7.5 BRANCH AND BOUND METHOD

The cutting plane method described in the last two sections is one way of exploring the integer lattice points (pure integer problems); the same approach may be extended to mixed integer problems. These problems are defined over a space in which variables can assume integer values in some directions and can vary continuously in the other directions. An alternative and by far the most viable method of exploring such solution spaces goes by the name of the Branch and Bound method.

Land and Doig [32] first introduced the concept of constructing a tree of subproblems and searching it in an ordered fashion. However, this approach has proved difficult to program. Dakin [14], Beale and Small [5], Driebeek [18] all developed branch and bound schemes on second generation computers; in this method a bifurcating tree of subproblems is constructed and these linear subproblems are suitably stored and subsequently solved.

A method very similar to the second type of tree search scheme is described in this section; for the purpose of generality a mixed integer problem is considered.

Consider the mixed integer linear programming problem

$$P_0\colon\ \text{Maximize}\ \{px + qy \mid Ax + By \leqslant b;\ x_j \geqslant 0, j \in N_1;\ y_j \geqslant 0 \text{ and } \equiv 0 \bmod (1); j \in N_2\} \qquad (7.24)$$

where p and x are n_1-vectors, $N_1 = 1, 2, \ldots, n_1$, q and y are n_2-vectors, $N_2 = 1, 2, \ldots, n_2$ A and B are $m \times n_1$ and $m \times n_2$ matrices respectively. The (all) continuous variable problem associated with P_0 is first solved. If in the optimal solution, one or more variables y_j turn out to be non-integer that is

$$y_j = \beta_j + f_j; \qquad \beta_j \equiv 0 \bmod (1), \qquad 0 < f_j < 1, \qquad (7.25)$$

one such variable is chosen to propose a dichotomy which generates two subproblems P_L and P_R†:

$$P_L\text{: Problem } P_0 \text{ with the added constraint } y_j \leqslant \beta_j \tag{7.26}$$

$$P_R\text{: Problem } P_0 \text{ with the added constraint } y_j \geqslant \beta_j + 1. \tag{7.27}$$

In general, this may be considered a typical step of a process whereby whenever the solution of an associated continuous LP problem fails to satisfy the integer conditions two subproblems of the type P_L and P_R are proposed. The subscripts L, R are indices identifying subproblems which may themselves be subsequently solved: a step commonly described as developing a node. This therefore leads to the development of a binary tree of problems, as shown in Fig. 7.2.

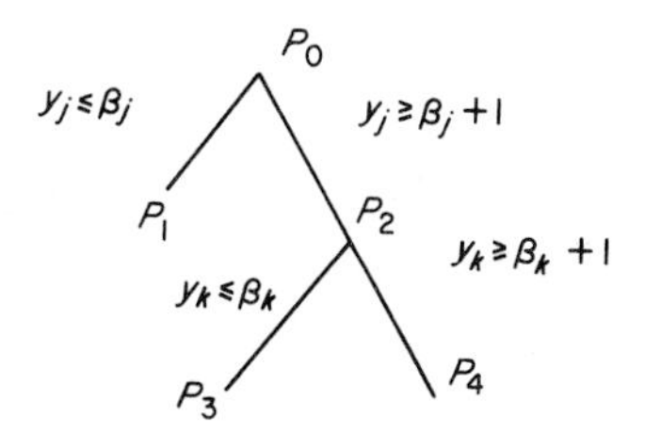

FIG. 7.2

Let C_i denote the Optimum value of the objective function of the continuous problem, corresponding to the ith subproblem, and I_i denote the value of the objective function at an optimum integer solution of the ith subproblem. Then the following relations are easily seen to be true:

$$I_N = \max\{I_L, I_R\}, \tag{7.28}$$

$$C_N \geqslant C_L, C_R, \tag{7.29}$$

where the subscript N identifies the parent problem from which the two subproblems of types L and R are generated. If H denotes the set of indices identifying all the terminal nodes at any stage of the development of the tree then

$$\max_{k \in H} C_k \geqslant I_0 \geqslant I_i,$$

where I_i is the value of the objective function of any intermediate integer solution and I_0 the optimum integer solution sought.

† Note the region $\beta_j < y_j < \beta_j + 1$ cannot contain an integer-feasible solution.

The steps of a general algorithm for the solution of a mixed integer (or a pure integer) linear programming problem may be stated as below.

(i) *Initial step*

Clear the integer solution marker: initialize max (I) [a location to store best integer solution found so far] to a large negative value. Solve the continuous *LP* problem P_0. If an optimum solution exists go to (v), otherwise go to (vii)

(ii) *Choose branching variable*

Choose a variable y_k ($y_k = \beta_k + f_k$; $\beta_k \equiv 0 \bmod (1)$, $0 < f_k < 1$) for branching add the two subproblems P_L, P_R to the list of subproblems and store the associated bases.

(iii) *Select subproblem (node) to solve*

If the list of subproblems is empty, go to (vii). Otherwise choose a subproblem P from the list such that the objective value of its parent minus the penalty† is greater than or equal to max (I) plus the tolerance.‡

(iv) *Solve subproblem*

Starting from the stored basis solve the subproblem using a simplex algorithm. If the algorithm used is of the dual type the subproblem may be discontinued should its objective function value become less than or equal to max (I); in this case go to (iii).

(v) *Analyse subproblem solution*

If the subproblem has no feasible solution go to (iii). If the objective function value is less than or equal to max (I) go to (iii). If the solution is integer (i.e., $y_j = 0 \bmod (1)$ for all $j \in N_2$ go to (vi). Otherwise go to (ii).

(vi) *Integer solution*

Set integer solution marker. The objective function value of this integer solution is greater than max (I), hence update max (I). Go to (iii).

† Penalty: Minimum value of the change in the objective function when a branching variable is set to its bound (it is considered later in this section).

‡ Tolerance: A positive real quantity specifying the range within which the optimum solution is sought.

(vii) *Exit*

If the integer solution marker is not set then no feasible (integer) solution exists to the problem. Otherwise output the best integer solution.

The size of the overall tree of subproblems which must be searched and solved is directly related to the computational effort necessary to solve the mixed integer problems. Strategies for the development of the tree derive directly from the two steps of the algorithm, namely, Choose Branching Variable (ii) and Select Sub-problems (iii). These strategies admit many variations and sophistications [38]; a standard strategy is described here for the purpose of simplicity.

If corresponding to a variable $y_k = \beta_k + f_k$, $k \in N_2 \cap B$† the constraint $y_k \geqslant \beta_k + 1$ is introduced, then the corresponding infeasibility will be $-(1 - f_k)$. Assume that the variable y_k is currently pivoted in the pth row of the optimal tableau. Then the change in the objective function in one dual iteration is,

$$\text{Pen}_u = \min_{j,\, \bar{a}_{pj} < 0} \bar{a}_{0j}(1 - f_k)/|\bar{a}_{pj}|, \qquad j \in N \tag{7.31}$$

where $\bar{a}_{0j}$ represents the reduced cost coefficient and $\bar{a}_{pj}$ updated row entry for the tableau under consideration. Pen_u is known as the up penalty associated with the variable y_k. Conversely, the penalty associated with introducing the constraint $y_k \leqslant \beta_k$ is,

$$\text{Pen}_d = \min_{j,\, \bar{a}_{pj} > 0} \bar{a}_{0j} f_k / \bar{a}_{pj}, \qquad j \in N \tag{7.32}$$

and this is known as the down penalty. Assume that the subproblems are stored and searched by a last in first out scheme then a tree development strategy due to Beale [5] and Driebeek [18] and now standardized may be stated as: "branch on the variable with the highest penalty and develop the node in the other direction".

7.6 AN EXAMPLE ILLUSTRATING THE BRANCH AND BOUND METHOD

Consider the problems stated in Section 4, (7.20); two further non binding upper bound constraints:

$$\left.\begin{array}{l} x_1 \leqslant M \\ x_2 \leqslant M \end{array}\right\} \text{ or } \left\{\begin{array}{l} x_1 + x_5 = M \\ x_2 + x_6 = M \end{array}\right\}, \quad x_5, x_6 \geqslant 0, \text{ integer}, \tag{7.32}$$

(M = 1000 a large positive integer),

† B is the set of indices of the basic variables. N is the set of indices of the nonbasic variables.

are added to introduce upper bound constraints explicitly. A very simple variable choice stategy (choose last non integer variable) is employed and the (nodes) subproblems are selected on last-in-first-out basis. The tableaux (set up in Tucker–Beale form) 7.8 to 7.18 illustrate how the exploration of the solution space corresponds with the exploration of the tree; also see Fig. 7.3.

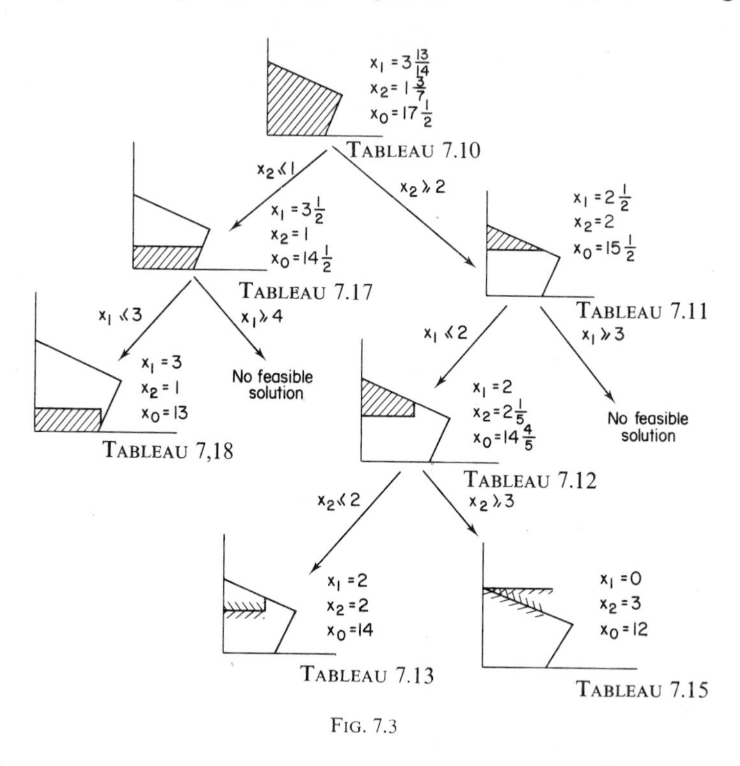

FIG. 7.3

In Tableau 7.10 the two penalties on the variable x_2, may be calculated (see (7.31), (7.32)) as,

$$\text{Pen}_u = (4/7 \times \tfrac{1}{2})/(1/7) = 2, \tag{7.34}$$

the penalty for setting the bound $x_2 \geqslant 2$; and

$$\text{Pen}_d = (3/7 \times 1)/(1/7) = 3, \tag{7.35}$$

the penalty for setting the bound $x_2 \leqslant 1$. Mitra [38] has reported some experiments concerning the usefulness of computing penalties, Benichou *et al.* [8] have reported employing pseudo cost criterion successfully for choosing variables.

Rewrite x_2 row as L.B.2 $| x_2 = -4/7 - 1/7(-x_4) + 1/7(-x_3)$

		$-x_1$	$-x_2$
x_0	0	-3	-4
x_3	15	2	(5)
x_4	5	2	-2
x_5	1000	1	0
x_6	1000	0	1
x_1	0	-1	0
x_2	0	0	-1

TABLEAU 7.8

		$-x_1$	$-x_2$
x_0	12	$-7/5$	$4/5$
x_3	0	0	-1
x_4	11	(14/5)	$2/5$
x_5	1000	1	0
x_6	997	$-2/5$	$-1/5$
x_1	0	-1	0
x_2	3	$2/5$	$1/5$

TABLEAU 7.9

		$-x_4$	$-x_3$
x_0	$35/2$	$1/2$	1
x_3	0	0	-1
x_4	0	-1	0
x_5	$996^1/14$	$-5/14$	$-1/7$
x_6	$998^4/7$	$1/7$	$-1/7$
x_1	$55/14$	$5/14$	$1/7$
x_2	$10/7$	$-1/7$	$1/7$

TABLEAU 7.10

Continuous optimum solution
Branch on variable x_2 and and store $x_2 \leqslant 1$ and explore $x_2 \geqslant 2$ TABLEAU 7.11

		pivot ↘ $-x_2$	$-x_3$
x_0	$31/2$	$7/2$	$3/2$
x_3	0	0	-1
x_4	4	-7	-1
x_5	$997^1/2$	$-5/2$	$-1/2$
x_6	998	1	0
x_1	$5/2$	$5/2$	$1/2$
LB2→ x_2	0	-1	0

TABLEAU 7.11

$x_1 \leqslant 2$
i.e. infeasibility row
$x_5 = -1/2 - 5/2(-x_2) - 1/2(x_3)$

Pivot

Now branch on x_1 and $x_1 \geqslant 3$ is not feasible.
So explore $x_1 \leqslant 2$

		$-x_5$	$-x_3$
x_0	$74/5$	$7/5$	$4/5$
x_3	0	0	-1
x_4	$27/5$	$-14/5$	$2/5$
x_5	0	-1	0
x_6	$997^4/5$	$2/5$	$-1/5$
x_1	2	1	0
LB2 x_2	$1/5$	$-2/5$	$1/5$

TABLEAU 7.12

$x_6 = -1/5 + 2/5(-x_5) - 1/5(-x_3)$

Pivot

List $x_2 \geqslant 3$ and explore $x_2 \leqslant 2$

		$-x_5$	$-x_6$
x_0	14	3	4
x_3	1	-2	-5
x_4	5	-2	2
x_5	0	-1	0
x_6	0	0	-1
x_1	2	1	0
LB2 x_2	0	0	1

TABLEAU 7.13

UB = 2

Tableau 7.13 is integer feasible so its objective function value 14 is taken as the value of the currently best integer solution. Extract the last problem from the list (Tableau 7.12; set lower bound $x_2 \geq 3$)

		$-x_5$	$-x_3$
x_0	74/5	7/5	4/5
x_3	0	0	-1
x_4	27/5	$-14/5$	2/5
x_5	0	-1	0
x_6	$997^4/5$	2/5	$-1/5$
x_1	2	1	0
LB $\rightarrow$ 3 x_2	$-4/5$	$-2/5$	1/5

TABLEAU 7.14

		$-x_4$	$-x_3$	
x_0	35/2	1/2	1	
x_3	0	-0	-1	
x_4	0	-1	0	
x_5	$996^1/14$	$-5/14$	$-1/7$	
x_6	$-3/7$	1/7	$(-1/7)$	UB = 1
x_1	55/14	5/14	1/7	
x_2	10/7	$-1/7$	1/7	

TABLEAU 7.16

		$-x_2$	$-x_3$
x_0	12	7/2	3/2
x_3	0	0	-1
x_4	11	-7	-1
x_5	2	1	0
x_6	997	1	0
x_1	0	5/2	1/2
LB3 $\rightarrow$ x_2	0	-1	0

TABLEAU 7.15

Integer feasible but worse than the best solution. Extract another problem from the list, Tableau 7.10 set $x_2 \leq 1$.

		$-x_4$	$-x_6$
x_0	29/2	3/2	7
x_3	3	-1	-7
x_4	0	-1	0
x_5	$996\frac{1}{2}$	$-1/2$	-1
x_6	0	0	-1
x_1	7/2	1/2	1
x_2	1	0	1

TABLEAU 7.17

$x_5 = -1/2 - 1/2 (-x_4) - (-x_6)$

Add $x_1 \leq 1$ to Tableau 7.17.

UB = 1

$x_1 \geq 2$ leads to no feasible solution.

		$-x_5$	$-x_6$	
x_0	13	3	4	
x_3	4	-2	-5	
x_4	1	-2	2	
x_5	0	-1	0	UB = 3
x_6	0	0	-1	
x_1	3	1	0	
x_2	1	0	1	

TABLEAU 7.18

Less than the currently best integer solution.

No problem left in the list; the search is completed.

EXERCISES

7.1 Describe a cutting plane method of pure integer programming and apply it to solve the problem

Maximize $\frac{1}{4}x_1 + x_2$

subject to $x_1/2 + x_2 \leqslant 7/4.$

$x_1 + 3x_2/10 \leqslant 7/5,$

and $x_1, x_2 \geqslant 0,$ and integer.

Illustrate the solution process graphically.

7.2 For an integer linear programming problem involving three variables and three constraints the optimum solution to the continuous linear programming problem is contained in the tableau below.

		$-x_5$	$-x_2$	$-x_6$
x_0	14	10/4	6/4	3
x_1	2/4	1/4	$-3/4$	0
x_3	18/4	3/4	$-1/4$	1
x_4	30/4	1/4	9/4	1

Carry out one cutting plane step towards obtaining an integer solution to this problem. From first principles deduce the equation of the cut which is introduced in this step. To solve an integer programming problem why is it most appropriate to use the Tucker–Beale form?

7.3 Apply the theory of Gomory's cutting plane method to derive stronger penalties on a branching variable than that obtained by the straight forward application of the duality theory.

CHAPTER 8

Formulating Mathematical Programming Models: Linear Programming, Integer Programming, and Nonlinear Programming by Extending Linear Programming Techniques

KEYWORDS: Distribution Models, Multiplant Production-cum-Distribution Model, Distribution Constraint, Material Balance Equation, Technical Constraint, Blending Constraint, Set up Cost, Discrete Function, Piecewise Linear Function, Separable Programming, Special Ordered Set, Chance Constrained Programming.

8.1 THE COMPONENTS OF A MATHEMATICAL PROGRAMMING MODEL

It is difficult to put in words the rules or the concepts involved in formulating models in the mathematical programming format. The task (and the mental activity involved) may be in many ways compared to that of solving a cross-word or putting together a jigsaw puzzle. Seemingly, the common denominator is that once the pieces (the variables and the constraints) are identified the overall shape emerges as a natural consequence. To formulate a mathematical programming problem, therefore the first task is to establish the the variables or the activities in which the problem is defined, and the restrictions (or constraints) which may be stated in these variables. It is then appropriate to look for the mathematical relationships which connect these variables and express the objective and the constraint functions.

Variables

The nature of the variables or the activities depend on the context of the models and may be best illustrated by examples.

Production planning. The quality X_{pm} of a certain product p that is produced on a machine m.

Distribution. The quantity X_{prn} of a product p that is shipped from a source r to a sink n.

Inventory control. The quantity X_{pt} of a product p that is kept as closing stock (inventory) at the end of time period t.

Investment problem. Whether one should invest in project p at the beginning of the time period $t(Y_{pt} = 1)$ or not invest $(Y_{pt} = 0)$ in this project at that time period may be expressed by the variable Y_{pt} which is a zero-one variable.

The solution to a mathematical programming problem is of course expressed as a vector of solution values of such variables. In a large real life model variables of different types may appear together: thus in a production-cum-distribution model both X_{pm} and X_{prn} variables may occur. At this stage of identifying the variables it is necessary to choose the appropriate dimensions (viz: thousands of gallons, or number of items etc.,) for each variable type.

Constraints.

The variables may assume values which are of necessity restricted. The restrictions may apply to individual variables or to each of the many linear forms of these variables or to nonlinear functions of these variables; the restrictions may be of one of the two types equalities or inequalities.

Bounding constraints.† In general, variables are non-negative and hence have a lower bound of zero. These may have a finite but nonzero lower bound. The individual variable may also have finite upper bound. The variable in this case is said to be bounded.

Generalized upper bound constraints. Instead of a single variable possessing an upper and a lower bound a group of variables may be summed up and constrained to be equal to or less than equal to a right hand side value. Such constraints are said to exhibit Generalized Upper Bound (GUB) structure.

Distribution constraints. The quantities X_{sw} of a product which arrive from different sources $s = 1, 2, \dots S$, say factories, must be less than or equal to

† Algorithms which take into account simple and generalized upper bounds are discussed in Chapter 5.

the handling capacity U_w of the warehouse, hence the relation

$$X_{1w} + X_{2w} + \ldots + X_{Sw} \leqslant U_w \tag{8.1}$$

illustrates a typical inequality with GUB structure.

Material balance equation. The GUB structure in its most general form (8.1) does not require that the coefficients of all the variables which appear in such an equation should be $+1$; thus a -1 coefficient is also admitted. Consider the material balance equation,

$$X_0 + X_p - X_c = D, \tag{8.2}$$

where X_0 is the opening inventory, X_p the quantity produced and X_c the closing inventory and D the customer demand. A number of exclusive equations (i.e. variables in one such equation does not appear in another such equation) of the type (8.2) also lead to a GUB structure. In identifying the GUB constraints in a model it is necessary to ensure that the exclusiveness condition above is satisfied and it is advantageous to find these rows in a way that their number is a maximum.

In general the constraints are not necessarily structured as explained above. Contexts in which constraints of other types arise are discussed in what follows.

Technical constraints. In process models the pressure of a certain vessel may not exceed an upper limit. The effluent of a furnace or an engine may not contain sulphur or lead above a certain part per million etc.

Blending constraint. The calorific value of a mixed feed or the octane number of a blended gaoline must be equal to or greater than equal to a specified amount.

8.2 MULTIPLANT MULTIPRODUCT PRODUCTION-CUM-DISTRIBUTION MODEL

A production planning problem involving more than one plant and more than one product may be stated as follows. At the two plants A, B situated at two different geographical locations there are demands DAP1, DAP2, DBP1, DBP2 of two different products P_1 and P_2. At A there are machines I, J, K, and at B there are machines L, M . HI, HJ, . . . , HM, are the total number of days that each of these machines are available in each month. The products may be transported between the plants after manufacture. Numerical information concerning some of the elements of the problem matrix such as the number of units of product P1 produced in a day, the cost of production, the cost of transport etc. are supplied in the Tables 8.1, 8.2. The situation is displayed in Fig. 8.1.

To formulate the problem of finding the operating schedule of the machines in each plant and also the interplant distribution of the products, all at a minimum cost, it is necessary to introduce the variables: XAIP1, XAIP2,..., XAKP2, the production variables for the machines at Plant A, XBLP1 ..., XBMP2, the production variables for the machines at Plant B, QABP1,...,

TABLE 8.1

Plants	Plant A Machines at this plant						Plant B Machines at this plant			
Machines	I		J		K		L		M	
Products	P1	P2	P1	P2	P1	P2	P1	P2	P1	P2
Cost/day	100	102	104	106	98	104	102	105	103	106
Prodn/day	40	35	42	38	40	37	41	37	42	40
Availability in days	30		28		24		26		28	

TABLE 8.2

Products	Product P1		Product P2	
Plants	A	B	A	B
Demands	1200	800	1500	1100
Transport cost/unit	From A to B = 4 From B to A = 4		From A to B = 3 From B to A = 4	

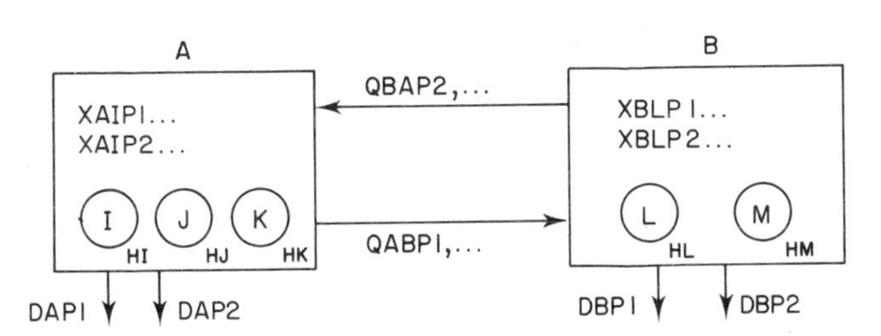

FIG. 8.1

QBAP2, the variables representing the transportation activities between the plants A, B. XAIP1 stands for number of days machine I at Plant A works to produce the product P1. QABP1 stands for the quantity P1 transhipped from A to B; the mnemonics are so chosen (this is normal in LP modelling) that

Constraints \ Variables	X A I P 1	X A I P 2	X A J P 1	X A J P 2	X A K P 1	X A K P 2	Q A B P 1	Q A B P 2	Q B A P 1	Q B A P 2	X B L P 1	X B L P 2	X B M P 1	X B M P 2	Nature of relation	Right hand side value
OP cost Operating cost	100	102	104	106	98	104	4	3	4	4	102	105	103	106	Free	—
DAP1 Demand of P1 at A	40		42		40		−1		1						=	1200
DAP2 Demand of P2 at A		35		38		37		−1		1					=	1500
HI Availability of M/C I	1	1													⩽	30
HJ Availability of M/C J			1	1											⩽	28
HK Availability of M/C K					1	1									⩽	24
DBP1 Demand of P1 at B							1		−1		41		42		=	800
DBP2 Demand of P2 at B								1		−1		37		40	=	1100
HL Availability of M/C L											1	1			⩽	26
HM Availability of M/C M													1	1	⩽	28
	Plant A model						Transportation model				Plant B model					

TABLEAU 8.3

the representation of the other variables naturally follows. Using these variables the problem may be set up as shown in Tableau 8.3.
The following observations concerning the model may now be made.

Number of variables and constraints: for a model with 5 plants and 20 products, and each plant capable of shipping goods to any of the four other plants, there will be a total of 20×4 outbound transhipment variables; taking all the five plants into account the total number of variables will be $5 \times 4 \times 20 = 400$. Taking production variables into account and assuming that there is an average of three machines per plant the number of production variables per plant is then $20 \times 3 = 60$ and the total number of production variables becomes $60 \times 5 = 300$. The capacity constraints for the model are three per plant, and demand constraints 20 per plant; hence the total number of constraints in the model is $23 \times 5 = 115$. This rectangular structure 115×700 is typical of this model. The fact it has so many variables outlines different strategies which can be followed; each column of transhipment or production variable represents transportation or production decisions.

Inclusion of the transhipment variables. Normally linear programs are defined in non-negative variables. It is natural to propose that the transhipment variables be defined free. For instance the variable QABP1 may be allowed to take negative value and the negative value represent the variable QBAP1. However, the cost of flow in each case is positive, and may even be unequal, hence the number of transhipment variables cannot be halved in this fashion.

Generalized upper bound (GUB) structure of the model

All the machine capacity constraints make up typical generalized upper bound (cf. Chapter 5) constraints of the model. However, the percentage of these rows depend on the number of products in the model. In general, if more than 15% of the rows are GUB then it is worth solving it by GUB algorithm in which case the transhipment and the production variables should be presented in two separate groups in the input data defining the model.

Multitime period models. In an industry with seasonal change in demand and with some seasonality of supply as well, such models should be constructed on the basis of a few distinct time periods per year. In this case the number of constraints multiply by the number of time periods and so do the number of variables. A few extra variables representing warehousing and stocking of each product per time period is introduced. In this case a staircase structure of the constraint relation emerges, although within a block of this structure the model remains more or less the same.

8.3 THE FIXED CHARGE PROBLEM

Consider the nonlinear programming problem

Minimize $f(x) = \phi_1(x_1) + \phi_2(x_2) + \ldots + \phi_n(x_n)$

subject to
$$a_{11}x_1 + a_{12}x_2 + \ldots + a_{1n}x_n = b_1 \qquad (8.3)$$
$$a_{m1}x_1 + a_{m2}x_2 + \ldots + a_{mn}x_n = b_m$$

and
$$x_1, x_2 \ldots x_n \geqslant 0.$$

Where $\phi_j(x_j) = c_jx_j + d_j\delta_j,\ d_j > 0$

$\delta_j = 0 \text{ or } 1$

and further
$$\left.\begin{array}{ll}\delta_j = 0 & \text{if } x_j = 0 \\ \delta_j = 1 & \text{if } x_j > 0\end{array}\right\} \quad j = 1, 2, \ldots n \qquad (8.4)$$

A typical function $\phi_j(x_j)$ is shown in the graphical form in Fig. 8.2.

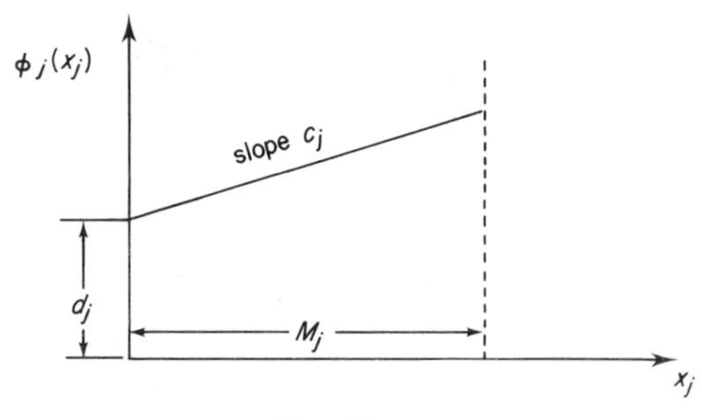

FIG. 8.2

For each continuous variable x_j introduce M_j as the corresponding realistic upper bound. The problem as stated in (8.3) is a non-linear problem; it may now be restated as a mixed integer linear programming problem in the following way,

$$\text{Minimize } f(x) = \sum_{j=1}^{n} (c_jx_j + d_j\delta_j)$$

subject to
$$a_{11}x_1 + a_{12}x_2 + \ldots + a_{1n}x_n = b_1 \qquad (8.5)$$
$$\vdots \qquad\qquad \vdots \qquad \vdots$$
$$a_{m1}x_1 + a_{m2}x_2 + \ldots + a_{mn}x_n = b_m$$
$$x_1 \ldots x_n \geqslant 0$$

further
$$\left.\begin{aligned} & x_j \leqslant M_j\delta_j \\ & 0 \leqslant \delta_j \leqslant 1, \text{and } \delta_j \text{ integer} \end{aligned}\right\} \text{for } j = 1, 2, \ldots, n.$$

The solution to the fixed charge problem has interesting properties [41], (8.1) which may be exploited in designing special purpose algorithms for their solution. It may be proved [27] that a local optimum solution in this case (minimizing a concave function subject to linear constraints) must be a basic feasible solution; hence the global optimum solution must also be a basic solution.

8.4 THE CAPITAL BUDGETING PROBLEM

A firm has a capital D to spend on $j = 1, 2, \ldots, n$ different projects. The profit associated with project j is p_j (it is more appropriate to consider this to represent the present worth of future return). d_j is the cost of investing in the project j. To arrive at an investment decision whereby the present worth of future profits is maximized the problem may be formulated as,

$$\text{Maximize } P = p_1\delta_1 + p_2\delta_2 + \ldots + p_n\delta_n \tag{8.6}$$

$$\text{subject to } d_1\delta_1 + d_2\delta_2 + \ldots + d_n\delta_n \leqslant D$$

$$0 \leqslant \delta_j \leqslant 1 \text{ and } \delta_j \text{ integer for all } j.$$

Here $\delta_j = 1$ represents that there should be an investment in project j and $\delta_j = 0$ represent the contrary.

There may be further restrictions whereby if the kth project is implemented then the lth project must also be implemented and vice versa. Such constraints may be expressed as

$$\delta_k - \delta_1 = 0. \tag{8.7}$$

Many other linear restrictions may be added to make the model represent more realistic real life situations.

8.5 INTEGER PROGRAMMING FORMULATION OF NON-LINEAR FUNCTIONS OF ONE VARIABLE

Consider the function $\phi(x_j)$ defined by the points $P_1, P_2, \ldots, P_5$ as shown in Fig. 8.3. Given the tabulated ordinates (a_k, b_k); $k = 1, 2, \ldots, 5$. the function

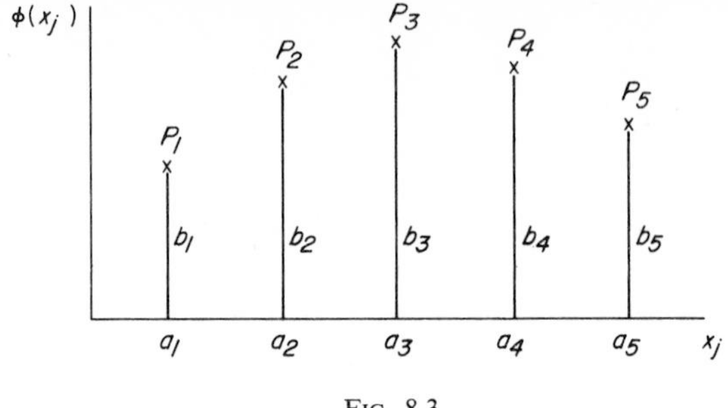

FIG. 8.3.

$\phi(x_j)$ may be represented as

$$\begin{aligned} \phi(x_j) &= b_1\lambda_1 + b_2\lambda_2 + \ldots + b_5\lambda_5 \\ x_j &= a_1\lambda_1 + a_2\lambda_2 + \ldots + a_5\lambda_5 \\ 1 &= \lambda_1 + \lambda_2 + \ldots + \lambda_5 \\ 0 \leqslant \lambda_k &\leqslant 1 \text{ and } \lambda_k \text{ integer, } k = 1, 2, \ldots, 5 \end{aligned} \tag{8.8}$$

The combinatorial nature of this discrete function should be obvious; for it can assume one of the five values weighted by λ_k only one of which can be unit at any one time.

Piecewise linear but continuous non-linear functions call for the introduction of a few more variables. Consider the function $\phi(x_j)$ shown in Fig. 8.4 which is similar to that shown in Fig. 8.3 but is further continuous i.e., defined by the line segments joining $P_1 - P_2, P_2 - P_3$, etc.

To formulate this by mixed integer programming introduce continuous variables $\lambda_k, 0 \leqslant \lambda_k \leqslant 1$ for $k = 1, 2, \ldots, 5$ and 0–1 integer variables δ_l, $l = 1, 2, \ldots 4$. Then the function $\phi(x_j)$ may be represented by the set of

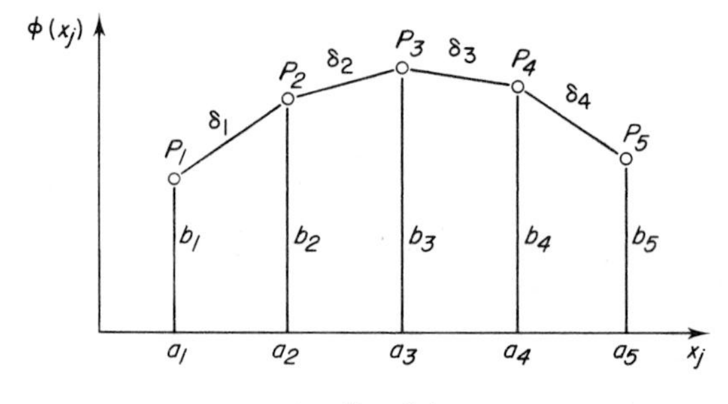

FIG. 8.4.

relations,

$$
\begin{aligned}
\phi(x_j) &= b_1\lambda_1 + b_2\lambda_2 + \ldots + b_5\lambda_5 \\
x_j &= a_1\lambda_1 + a_2\lambda_2 + \ldots + a_5\lambda_5 \\
1 &= \lambda_1 + \lambda_2 + \ldots + \lambda_5, \\
\lambda_1 &\leqslant \delta_1 \\
\lambda_2 &\leqslant \delta_1 + \delta_2 \\
\lambda_3 &\leqslant \delta_2 + \delta_3 \\
\lambda_4 &\leqslant \delta_3 + \delta_4 \\
\lambda_5 &\leqslant \delta_4 \\
&\delta_1 + \delta_2 + \delta_3 + \delta_4 = 1 \\
&0 \leqslant \lambda_k \leqslant 1, k = 1, \ldots 5, \\
&\delta_l = 0 \text{ or } 1, \quad l = 1, 2, 3, 4.
\end{aligned}
\tag{8.9}
$$

In the formulation above observe that for any δ_j taking the value 1 ensures that the point lying on the corresponding segment is a linear combination of the end points, i.e., only λ_j, λ_{j+1} can have positive value.

8.6 PIECEWISE LINEAR REPRESENTATION OF NON-LINEAR FUNCTION BY SLOPE AND INTERCEPT

An alternative representation of the piecewise linear function was first furnished by Healey [29].

Consider the function $\phi(x_j)$ Fig. 8.5 of the variable x_j and represented by the straight line segments between $P_1 - P_2$, $P_2 - P_3$ etc. Let $L_1, L_2, \ldots, L_4$ be the lower bounds and $U_1, U_2, \ldots, U_4$ the upper bounds on the values of x_j lying in the argument of each of the segments that is $\phi(x_j) = a_i + b_i x_j$,

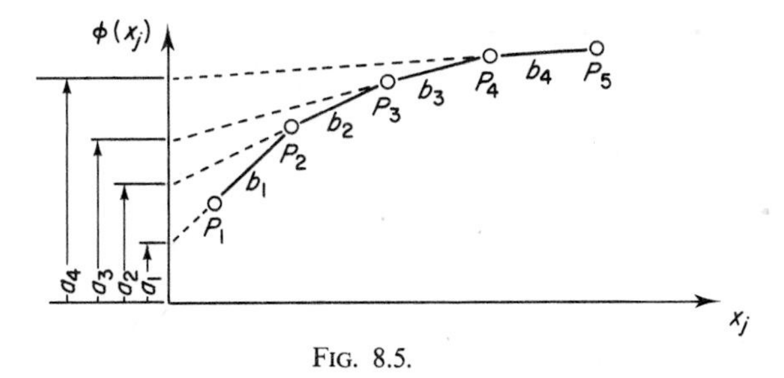

FIG. 8.5.

$L_i \leqslant x_j \leqslant U_i$. It is not necessary to postulate $U_1 = L_2$ etc., although this is so in the present case. Let the straight line segments themselves be given by the equations $a_i + b_i x_j$; $i = 1, 2, \ldots, 4$ where a_i represents the intercept on the $\phi(x_j)$ axis and b_i the slope of the line. Define a set of four integer variables δ_i, $i = 1, 2, \ldots, 4$ for the four segments, then the function may be represented as

$$\phi(x_j) = \sum_{i=1}^{4} \delta_i(a_i + b_i x_j) \tag{8.10}$$

and

$$\sum_{i=1}^{4} \delta_i = 1. \tag{8.11}$$

From (8.10) it follows that,

$$\begin{aligned} \phi(x_j) &= \sum_{i=1}^{4} \delta_i(a_i + b_i L_i + b_i[x_j - L_i]) \\ &= \sum_{i=1}^{4} \delta_i(a_i + b_i L_i) + \sum_{i=1}^{4} b_i w_i, \end{aligned} \tag{8.12}$$

where

$$w_i = \delta_i(x_j - L_i). \tag{8.13}$$

For $\delta_i = 1$, it is further necessary to have x_j to be between $[L_i, U_i]$ which implies that for $\delta_i = 1$, w_i must be in the range $[0, U_i - L_i]$. Sum both sides of (8.13) over $i = 1, 2, \ldots, 4$. The set of constraints for the representation of $\phi(x_j)$ can be expressed as,

$$\sum_{i=1}^{4} w_i + \sum_{i=1}^{4} \delta_i L_i = \sum_{i=1}^{4} \delta_i x_j = x_j \tag{8.14}$$

$$\phi(x_j) = \sum_{i=1}^{4} \delta_i(a_i + b_i L_i) + \sum_{i=1}^{4} b_i w_i \tag{8.15}$$

$$1 = \sum_{i=1}^{4} \delta_i$$

$$\delta_i(U_i - L_i) \geqslant w_i, \qquad i = 1, 2, \ldots, 4$$

and $\delta_i = 0$–1 integer.

The above set of constraints therefore provides a piecewise linear representation of the nonlinear function $\phi(x_j)$.

8.7 SEPARABLE PROGRAMMING AND SPECIAL ORDERED SETS OF TYPE 1 AND TYPE 2

If a non-linear function of many variables can be expressed as a sum of linear and non-linear functions of one variable at a time then the non-linear function is said to be variable separable. C. Miller [36] introduced separable programming: a means of formulating programming problems involving such non-linear functions and appropriately modifying the simplex algorithm for the solution of such problems. Consider a variable separable cost function,

$$f(x) = \phi_1(x_1) + \phi_2(x_2) + \ldots + \phi_n(x_n) \tag{8.16}$$

which must be minimised subject to some linear constraints,

$$\begin{aligned} a_{11}x_1 + a_{12}x_2 + \ldots + a_{1n}x_n &= b_1 \\ a_{m1}x_1 + a_{m2}x_2 + \ldots + a_{mn}x_n &= b_m \end{aligned} \tag{8.17}$$

and $$x_1 \ldots x_n \geqslant 0.$$

Then any one of these functions $\phi_j(x_j)$, Fig. 8.6 may be represented by the

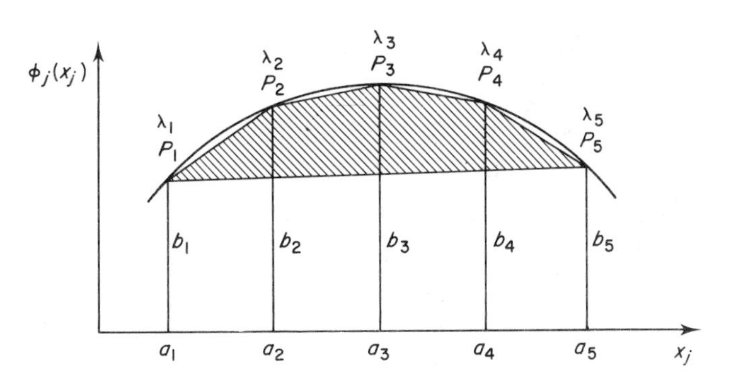

FIG. 8.6.

following set of equations,

$$\begin{aligned} \phi_j(x_j) &= b_1\lambda_1 + b_2\lambda_2 + \ldots + b_5\lambda_5 \\ x_j &= a_1\lambda_1 + a_2\lambda_2 + \ldots + a_5\lambda_5 \text{ reference row} \\ 1 &= \lambda_1 + \lambda_2 + \ldots + \lambda_5 \text{ convexity row} \end{aligned} \tag{8.18}$$

$$0 \leqslant \lambda_1 \ldots \lambda_5 \leqslant 1$$

plus the added condition that only two adjacent λs can be positive at any one time; $\lambda_1 \ldots \lambda_5$ are introduced to linearize the function. This is a piecewise linear approximation to the function $\phi_j(x_j)$, where 5 points on the curve are first plotted and their x_j and function value $\phi_j(x_j)$ e.g. (a_i, b_i) are noted for these points. The set of weighting variables λs are called the special variables and the two rows in the formulation are called the reference row and the convexity row respectively.

Note that if the adjacency condition is not imposed on the λs then the convex region showed by the shaded area is represented by this formulation.

Finally note that either the objective function or the constraint function of a programming problem may be linearized in this way.

Computational aspects of separable programming

If the constraint set of the programming problem is convex or the objective function of the minimizing problem is convex then it is not necessary to resort to special algorithms to satisfy the adjacency conditions. However, if the objective function is not convex then the following modification needs to be carried out in the PRICE step of simplex method (in this step one decides which variable to bring into the basis depending on the negativity of its reduced cost coefficient). The necessary modifications of the usual simplex steps are as follows,

if no special variable from a set is in the basis then all the non basic special variables are priced:

if only one special variable from a set is in the basis then its two adjacent variables are priced;

if two adjacent special variables are in the basis then no other variable from this set is priced.

However, there are two further extensions to such a scheme. It may be that such a restricted PRICE scheme leads to a no-feasible solution in an irrelevant way. To get round this, one solves the problem without separable conditions and may obtain an illegal solution. One can then carry out an interpolation procedure whereby the corresponding feasible (legal) solution for the current $\phi_j(x_j)$ and/or x_j value is constructed. Subsequent application of separable PRICE will lead to a local optimum solution.

It is also possible to make the piecewise linear approximation finer in the region of interest of the function. Both these facilities may be found in a commercial LP system such as UMPIRE [49].

Special ordered sets of Type 1 *and Type* 2.

It follows from the definition of the separable functions and the earlier integer programming formulation of these functions that there are two cases of interest:

the function under consideration is defined by a discrete set of points corresponding to a set of discrete values of x_j;

the piecewise linear interpolates given by the lines joining points on the function replaces the function.

If in the set of constraints (8.18) the adjacency condition is replaced by $\lambda_j = 0$ or 1 then case 1 is obtained; this is called special ordered set of type 1. Case 2 is of course the straight forward separable programming formulation and it is called special ordered set of Type 2. These sets of variables are called special ordered because the ordering $a_1 \leqslant a_2 \leqslant a_3$ etc. is implied in their formulation. The terminology is due to Beale and Tomlin [6], who have devised a branch and bound scheme for the global optimal solution of both these types of problems.

8.8 SOLUTION OF CHANCE CONSTRAINED PROGRAMMING PROBLEMS BY THE USE OF SEPARABLE PROGRAMMING

In this section an application model which illustrates the "mathematical programming" approach to a class of "statistical decision problems", is illustrated. One popular name for this model is "deterministic non-linear programming equivalent for the stochastic linear programming problem."

Given the triplet (A, b, c) where A is an $m \times n$ matrix and b an m-vector and c and n-vector of coefficients, the linear programming problem

$$\begin{aligned} &\text{Minimize} && c'x \\ &\text{subject to} && Ax \geqslant b \\ & && x \geqslant 0, \end{aligned} \tag{8.19}$$

may be stated with a vector x of n unknown variables restricted to non-negative values. This formulation of the model and the interpretation of its results are carried out under the assumption that the triplet (A, b, c) is wholly made up of constants. Now assume that one or more members of this triplet is subject to error; a variety of approaches may be employed to analyse the resulting model. For instance one may apply "sensitivity analyses" to a part or all of the entire system of relations. This may be done by suitable parametrization techniques or by applying direct dual relations, and systematically exhibiting the optimal patterns of variations.

The chance constrained formulation of the linear programming problem stated in (8.19) extends the deterministic problem in the following way:

$$\text{Minimize} \quad \sum_{j=1}^{n} c_j x_j$$

subject to $$\Pr\left[\sum_{j=1}^{n} \tilde{a}_{ij}x_j \geqslant b_i\right] \geqslant \alpha_i = 1 - \varepsilon_i, \qquad i = 1, 2, \ldots, m, \qquad (8.20)$$

$$x_j \geqslant 0, j = 1, 2, \ldots, n,$$

where some or all the coefficients of a_{ij} are random variables with mean $\tilde{a}_{ij}$ and normal distributions, and $\alpha_i = 1 - \varepsilon_i$ are the specified probabilities with which the m constraints may be satisfied for $i = 1, 2, \ldots m$.
A study due to Van-de-Panne [50] is now considered to elucidate one such model. Table 8.3 specifies the constituents of the ingredients viz Barley etc. with which a cattle feed is blended. Of the two nutritive requirements Fat and Protein the latter is more important. The mean of the protein content varies from sample to sample hence in addition to the mean the variance is also noted for the protein contents of all the ingredients.

TABLE 8.3

		Protein content mean/variance $\tilde{a}_{ij}/\sigma^2_{ij}$ $i = 1$	Fat content mean $\tilde{a}_{ij}$ $i = 2$	Cost per ton in guilders
Barley	x_1	12·0/0·28	2·3	24·55
Oats	x_2	11·9/0·19	5·6	26·75
Sesame flakes	x_3	41·8/20·5	11·1	39·00
Groundnut meal	x_4	52·1/0·62	1·3	40·50

The LP formulation of the cattle feed problem would be to find the feed mix, viz $x_1, x_2, \ldots, x_4$? such that the cost of the feed is minimized subject to minimum protein and fat contents of 21, and 5 units; i.e.

$$\begin{aligned} \text{Minimize} \quad & 24{\cdot}55x_1 + 26{\cdot}75x_2 + 39{\cdot}00x_3 + 40{\cdot}50x_4 \\ \text{subject to} \quad & 12{\cdot}0x_1 + 11{\cdot}9x_2 + 41{\cdot}8x_3 + 52{\cdot}1x_4 \geqslant 21 \qquad (8.21) \\ & 2{\cdot}3x_1 + 5{\cdot}6x_2 + 11{\cdot}1x_3 + 1{\cdot}3x_4 \geqslant 5 \\ & x_1 + x_2 + x_3 + x_4 = 1 \\ & x_1, x_2, \ldots, x_4 \geqslant 0. \end{aligned}$$

The optimal solution to this problem,

$$x_1 = 0{\cdot}6852, \qquad x_2 = 0{\cdot}0127, \qquad x_3 = 0{\cdot}3021, \qquad x_4 = 0{\cdot}0$$

and

$$\text{cost} = 28{\cdot}94 \text{ guilders/ton}$$

is obtained by the straightforward application of the simplex algorithm.

Assume now that the protein content of the raw materials used for one batch of the mixture is constant and variations of protein contents in inputs occur only for different batches of the mixture. If the protein content of the raw materials varies then using the mean values as in the linear programming problem one cannot be sure that a specific sample of the mixture satisfies the requirements. This difficulty can be dealt with by changing the requirement of the minimum protein content into a probabilistic one: viz. by requiring that the probability of a protein content in the mixture equal to or larger than a certain minimum should not be lower than a given level (say 0·95). If $f(x)$ denotes the protein content in a sample of the mixture which contains proportions of the raw materials given by the vector x, then the new constraint may be expressed as

$$\Pr\left[f(x) \geqslant b_1\right] \geqslant 1 - \varepsilon_1, \tag{8.22}$$

where $b_1 = 21$ is the protein level specified and $1 - \varepsilon_1$ is the minimum probability, ε_1 being the proportion of cases when a deficient protein level is allowed to occur. Let $\tilde{a}_{1j}$ be the expected protein content of the input j and σ_j^2 its variance and assume that there is no correlation for different inputs, the expected protein content of any mixture $(x_1 \ldots x_4)$ is $\sum_{j=1}^{4} \tilde{a}_{1j}x_j$, and its variance is $\sum_{j=1}^{4} \sigma_{1j}^2 x_j^2$. From (8.22) the following inequality may be derived

$$\sum \tilde{a}_{1j}x_j + \phi\left(\sum \sigma_{1j}^2 x_j^2\right)^{\frac{1}{2}} \geqslant b_1, \tag{8.23}$$

where ϕ is the normal deviate corresponding to the probability:

$$\varepsilon_1 = \frac{1}{\sqrt{2\pi}} \int_{-\infty}^{\phi} \exp\left(-x^2/2\right) dx. \tag{8.24}$$

The linear programming model can be now reformulated as the chance constrained problem,

$$\begin{aligned}
\text{Minimize} \quad & 24{\cdot}55x_1 + 26{\cdot}75x_2 + 39{\cdot}00x_3 + 40{\cdot}50x \\
\text{subject to} \quad & 12{\cdot}0x_1 + 11{\cdot}9x_2 + 41{\cdot}8x_3 + 52{\cdot}1x_4 \\
& + (-1{\cdot}645)(0{\cdot}28x_1^2 + 0{\cdot}19x_2^2 + 20{\cdot}5x_3^2 + 0{\cdot}62x_4^2)^{\frac{1}{2}} \geqslant 21 \\
& 2{\cdot}3x_1 + 5{\cdot}6x_2 + 11{\cdot}1x_3 + 1{\cdot}3x_4 \geqslant 5 \\
& x_1 + x_2 + x_3 + x_4 = 1 \\
& x_1, x_2, \ldots, x_4 \geqslant 0.
\end{aligned} \tag{8.25}$$

The non-linear programming problem as stated in (8.25) may be solved by separable programming technique for it can be expressed in the following variable separable form;

$$\begin{aligned}
\text{Minimize} \quad & 24{\cdot}55x_1 + 26{\cdot}75x_2 + 39{\cdot}00x_3 + 40{\cdot}50x_4 \\
\text{subject to} \quad & 12{\cdot}0x_1 + 11{\cdot}9x_2 + 41{\cdot}8x_3 + 52{\cdot}1x_4 - 1{\cdot}645y^{\frac{1}{2}} \geqslant 21 \\
& 0{\cdot}28x_1^2 + 0{\cdot}19x_2^2 + 20{\cdot}5x_3^2 + 0{\cdot}62x_4^2 = y \qquad (8.26) \\
& 2{\cdot}3x_1 + 5{\cdot}6x_2 + 11{\cdot}1x_3 + 1{\cdot}3x_4 \geqslant 5 \\
& x_1 + x_2 + x_3 + x_4 = 1 \\
& x_1, \ldots, x_4 \geqslant 0.
\end{aligned}$$

In this $y^{\frac{1}{2}}, x_1^2 \ldots x_4^2$ are all variable separable and hence may be linearized by methods explained in section 7 of this chapter. This problem was solved in this way using XDLA system to yield the solution $x_1 = 0{\cdot}636$, $x_2 = 0{\cdot}0$, $x_3 = 0{\cdot}313$, $x_4 = 0{\cdot}051$ which is the same as that obtained in [9], [50]. The model may be generalized to any number of raw materials (variables) and and any number of probabilistic constraints in addition to the linear ones, i.e.,

$$\begin{aligned}
\text{Minimize} \quad & F(x) = \sum_{j=1}^{n} c_j x_j \\
\text{subject to} \quad & f_i(x) = \sum_{j=1}^{n} \tilde{a}_{ij} x_j + \phi\Big(\sum_{i,j} v_{ij} x_i x_j\Big)^{\frac{1}{2}} \geqslant b_i \qquad (i = 1, 2, \ldots, p) \\
\text{and} \quad & \sum_{j=1}^{n} a_{ij} x_j \geqslant b_i \qquad (i = p + 1, \ldots, m) \qquad (8.27) \\
& x_1 \ldots x_n \geqslant 0,
\end{aligned}$$

where v_{ij} are the elements of the variance–covariance matrix V. This is a convex programming problem a proof of the convexity of this model is given in [50].

EXERCISES

8.1 Reformulate the problem of generation planning as discussed in Chapter 1, Section 2, as a multitime period model of 5 time periods.

8.2 Reformulate the problem of section 1 Chapter 8 as a multitime period model of 5 time periods. Introduce appropriate inventory carrying charge and assume that the inventory can be carried from one time period to the next.

8.3 Under normal working conditions a factory produces 100 units of a certain product in each of four consecutive time periods at costs which vary from period to period as shown in the table below.

Additional units can be produced by overtime working, the maximum quantity and the costs again being shown in the table, as also is the demand for the product in each of four time periods. Moreover, it is possible to hold up to 60 units of product in store from one period to the next at a cost of £1K per unit per period together with a handling cost of £2K per unit associated with getting it into and out of store. It is required to determine that production and storage schedule which will meet the stated demands over the four time periods at minimum cost. Formulate this as a linear programming problem.

Time period	Demand units	Normal prods. costs (£K/unit)	Overtime prods. capacity (units)	Overtime prods. costs (£K/unit)
1	135	6	60	8
2	70	4	65	6
3	125	8	65	10
4	195	9	60	11

8.4 For Exercises 8.1, 8.2, 8.3 estimate the number of rows, and columns in the models. In each of these cases also find out the number of GUB (Generalized Upper Bound) rows.

8.5 Reformulate the problem of optimal redundancy discussed in Chapter 1, Section 2, as
(i) mixed integer linear programming problem,
(ii) special ordered set problem of type 1.

8.6 Introducing 0–1 integer variables formulate the following constraints ((i) is an illustration) (ii), (iii), (iv),

(i) $x_1 + x_2 \leqslant 4$,
$x_1 \geqslant 1$ or $x_2 \geqslant 1$,
and $x_1, x_2 \geqslant 0$

is equivalently stated as

$$x_1 + x_2 \leqslant 4,$$
$$x_1 - 1 + (1 - \delta_1) \geqslant 0,$$
$$x_2 - 1 + (1 - \delta_2) \geqslant 0,$$
$$\delta_1 + \delta_2 = 1$$
$$\delta_1, \delta_2 \geqslant 0 \text{ and integer},$$
$$x_1, x_2 \geqslant 0$$

(ii) $x_1 + x_2 \leqslant 4$,
$x_1 \geqslant 3$, or $x_2 \geqslant 2$,
and $x_1, x_2 \geqslant 0$

(iii) $\begin{cases} x_1 + x_2 \leqslant 4, \\ x_2 \geqslant 2, \end{cases}$

or

$\begin{cases} x_1 \leqslant 3 \\ x_2 \leqslant 1, \end{cases}$

$x_1, x_2 \geqslant 0.$

(iv) $\begin{Bmatrix} x_1 \leqslant 2 \\ x_2 \geqslant 1 \end{Bmatrix}, \begin{Bmatrix} x_1 \leqslant 4 \\ x_2 \leqslant 2 \end{Bmatrix},$

$\begin{Bmatrix} x_1 \geqslant 3 \\ x_2 \leqslant 1 \end{Bmatrix}, \begin{Bmatrix} x_1 + 2x_2 \geqslant 6 \\ x_1 + 2x_2 \leqslant 8 \end{Bmatrix},$

$x_1, x_2 \geqslant 0$, and any two of these four constraints must be effective at any one time.

8.7 Formulate the following problem as a mixed integer linear programming problem or a special ordered set problem,

$$\begin{aligned} \text{Maximize } & x_1x_2 + x_2x_3, \\ \text{subject to } & x_1 + 2x_2 + 3x_3 \leqslant 6, \\ & 3x_1 + x_2 + 2x_3 \leqslant 8, \end{aligned}$$

and $\quad x_1, x_2, x_3 \geqslant 0.$

8.8 Reformulate the Fixed Change problem discussed in Section 3, Chapter 8, as a separable programming problem or a special ordered set problem.

CHAPTER 9

The General Mathematical Programming Problem: Lagrange and Kuhn-Tucker Multipliers

KEYWORDS: Lagrangean Function, Lagrange Multiplier, Kuhn–Tucker Multiplier, Complementarity Condition, Stationary Point.

9.1 GENERALIZED MATHEMATICAL PROGRAMS

The problem:

$$
\begin{array}{llll}
\text{Maximize} & f(x_1, x_2, \dots, x_n) & & \\
\text{subject to} & g_i(x_1, x_2, \dots, x_n) = b_i & i = 1, 2, \dots, p & (9.1) \\
& g_i(x_1, x_2, \dots, x_n) \leqslant b_i & i = p + 1, \dots, m & \\
\text{and} & x_1 \dots x_n \geqslant 0, & &
\end{array}
$$

may be considered as the most general statement of the constrained optimisation problem. In this statement no assumption is made concerning the properties (such as the continuity of the first derivative or that of the second derivative) of the objective or the m constraint functions. The problem has been considered in the classical literature [28], and in more recent works [1], [27]. In the late forties and early fifties Kuhn and Tucker [30] investigated this problem rigorously and later Fiacco and McCormick [21] put forward non-classical methods for its solution.

To study the methods discussed in the rest of this chapter and in the following chapter the reader is expected to have an understanding of the material presented in Appendix 1 of this book.

9.2 LAGRANGE MULTIPLIERS FOR OPTIMIZATION PROBLEMS WITH EQUALITY CONSTRAINTS

This classical technique was developed to derive the equations of constrained motion of particles, and has been applied to solve optimization problems with only equality constraints, viz.,

$$\begin{aligned} &\text{Maximize} \quad f(x) \\ &\text{subject to} \quad g_i(x) = b_i, i = 1, 2, \ldots, m. \end{aligned} \tag{9.2}$$

The derivation of the theory of this method is best illustrated by considering a two variable problem with a single constraint as shown below:

$$\begin{aligned} &\text{Maximize} \quad z = f(x_1, x_2) \\ &\text{subject to} \quad g(x_1, x_2) = b. \end{aligned} \tag{9.3}$$

Assume that $f, g \in c'$ i.e. their first partial derivatives exist, and further assume that

$$\frac{\partial g(x^*)}{\partial x_2} \neq 0$$

i.e., this partial derivative (the determinant of the Jacobian in case of more than one constraint) does not vanish at the optimum solution. Then by the implicit function theorem stated in A1.7 there exists an eps-neighbourhood of x_1^* where it is possible to solve $g(x_1, x_2) - b = 0$, explicitly for x_2 to obtain the relationship

$$x_2 = \phi(x_1). \tag{9.4}$$

Substituting for x_2 in $f(x_1, x_2)$ it follows that

$$z = h(x_1) = f(x_1, \phi(x_1)),$$

for

$$|x_1 - x_1^*| < \text{eps}. \tag{9.5}$$

If f takes on a relative maximum at x^* for x satisfying $g(x_1, x_2) = b$, it must be true that there exists a δ such that $0 < \delta < \text{eps}$ and $h(x_1) < h(x_1^*)$ for all x_1 in this neighbourhood; in other words the function $h(x_1)$ has an unconstrained relative maximum at x_1^*. For the function $\phi(x_1)$ the differentiation is obtained by applying the implicit function theorem and for the function $h(x_1)$ the rules for differentiating compound functions are applied. Hence the conditions for the maximum of $h(x_1)$ may be deduced as,

$$\frac{\mathrm{d}h}{\mathrm{d}x_1} = \frac{\partial f}{\partial x_1} + \frac{\partial f}{\partial x_2} \cdot \frac{\partial \phi}{\partial x_1} = \frac{\partial f}{\partial x_1} - \frac{\partial f}{\partial x_2}\left(\frac{\partial g/\partial x_1}{\partial g/\partial x_2}\right) = 0, \tag{9.6}$$

where

$$\frac{\partial \phi}{\partial x_1} = -\left(\frac{\partial g/\partial x_1}{\partial g/\partial x_2}\right).$$

If a new variable λ is introduced to express the ratio,

$$\lambda = \frac{\partial f/\partial x_2}{\partial g/\partial x_2}, \tag{9.7}$$

then (9.6) reduces to

$$\frac{\partial f}{\partial x_1} - \lambda \frac{\partial g}{\partial x_1} = 0, \tag{9.8}$$

and (9.7) may be rewritten

$$\frac{\partial f}{\partial x_2} - \lambda \frac{\partial g}{\partial x_2} = 0. \tag{9.9}$$

To these append the equality constraint of (9.3)

$$g(x_1, x_2) = b \tag{9.10}$$

then the relationships (9.8), (9.9), (9.10) express the necessary (but not sufficient) conditions for obtaining the constrained maximum of the problem stated in (9.3). Note there are three equations (9.8), (9.9), (9.10) and there are three variables x_1, x_2, λ which take on the values x_1^*, x_2^*, λ^* at the optimum solution and also satisfy these equations. A less rigorous approach states that to obtain the optimum solution of (9.2) it suffices to introduce m lagrange multipliers, $\lambda = (\lambda, \lambda_2, \ldots, \lambda_m)$ and investigate the stationary points of the Lagrangean function,

$$F(x, \lambda) = f(x_1, x_2, \ldots, x_n) - \sum_{i=1}^{m} \lambda_i [g_i(x_1, x_2, \ldots, x_n) - b_i]. \tag{9.11}$$

It is then a necessary condition that the optimum solution of (9.2) is contained in one of the stationary points of (9.11), and hence must satisfy the $n + m$ equations obtained by taking n partial derivatives in x and m partial derivatives in λ and setting these expressions equal to zero,

$$\frac{\partial F}{\partial x_j} = \frac{\partial f}{\partial x_j} - \sum_{i=1}^{m} \lambda_i \frac{\partial g_i}{\partial x_j} = 0 \qquad j = 1, 2, \ldots, n$$

and

$$\frac{\partial F}{\partial \lambda_i} = g_i(x_1, x_2, \ldots, x_n) - b_i = 0 \qquad i = 1, 2, \ldots, m. \tag{9.12}$$

Provided the rank of the Jacobian associated with the optimum solution is m, i.e., $r(J_x^*) = m$, it is possible to determine uniquely the optimum solution x^* and the associated multipliers λ^*.

An example

Consider the problem,

$$\text{Minimize} \quad f(x_1, x_2) = 6 - 6x_1 + 2x_1^2 - 2x_1x_2 + 2x_2^2 \tag{9.13}$$

$$\text{subject to} \quad x_1 + x_2 = 2.$$

To solve the problem set up the Lagrangean, (for a maximizing function)

$$F(x, \lambda) = -(6 - 6x_1 + 2x_1^2 - 2x_1x_2 + 2x_2^2) - \lambda(x_1 + x_2 - 2), \tag{9.14}$$

and take partial derivatives with respect to x_1, x_2, λ and set these to zero

$$6 - 4x_1 + 2x_2 - \lambda = 0$$

$$2x_1 - 4x_2 - \lambda = 0$$

$$x_1 + x_2 - 2 = 0 \tag{9.15}$$

This set of equations may be solved to obtain the optimum solution, $x_1 = 3/2$, $x_2 = \frac{1}{2}$, $\lambda = 1$, and $f(x) = \frac{1}{2}$ the optimum value of the objective. The Lagrangean $F(x, \lambda)$ has only one stationary point the necessary condition in (9.15) for this convex problem is also sufficient to guarantee global optimality.

9.3 INTERPRETATION OF THE LAGRANGE MULTIPLIER

The Lagrange multiplier λ has the same interpretation as that of shadow price in linear programming. It measures the rate of change of the objective function in the optimum solution for a small change in the right hand side value of the corresponding constraint. In the problem (9.3) let the Lagrangean $F(x, \lambda)$ be set up:

$$F(x, \lambda) = f(x_1, x_2) - \lambda[g(x_1, x_2) - b]. \tag{9.16}$$

Taking partial derivatives w.r.t. b it follows,

$$\frac{\partial F}{\partial b} = \lambda.$$

In the last section $\lambda = 1$ in the optimum solution, which implies that if the r.h.s. changes by $\frac{1}{2}$ then the optimum solution value changes by $\frac{1}{2} \times 1 = \frac{1}{2}$ which is a positive increment.

The Lagrange multiplier is not restricted to take a particular sign in the optimal solution value. Consider the general problem stated in (9.2) and for which the Lagrangean is expressed in (9.11). Introduce a function $g'_i(x) = -g_i(x)$ for all i such that (9.2) may be stated as

$$\begin{aligned} &\text{Maximize} && f(x) \\ &\text{subject to} && -g'_i(x) = b_i \quad i = 1, 2, \ldots, m. \end{aligned}$$

Then the corresponding Lagrangean stated in m new Lagrange multiplies λ'_i takes the form

$$F'(x, \lambda') = f(x) + \sum_{i=1}^{m} \lambda'_i(g'_i(x) + b_i),$$

Of course the λ'_i in this problem has the sign opposite to that of λ_i in (9.11), although the same stationary points in x variables are obtained. This proves that the Lagrange multipliers are not restricted in sign.

9.4 KUHN–TUCKER MULTIPLIERS FOR PROBLEMS WITH INEQUALITY CONSTRAINTS

Kuhn and Tucker [30] in 1951 extended the concept and the application of the Lagrange multiplier to problems involving inequality constraints.

Consider the problem,

$$\text{Maximize} \quad f(x)$$

subject to the constraints,

$$g_i(x) \leqslant b_i \qquad i = 1, 2, \ldots, m. \tag{9.17}$$

By investigating the saddle point property of the Lagrangean,

$$F(x, \lambda) = f(x) - \sum_{i=1}^{m} \lambda_i(g_i(x) - b_i) \tag{9.18}$$

Kuhn and Tucker deduced the conditions for optimality. For non-convex problems these are necessary conditions for local optimum, for convex problems the necessary conditions are also sufficient to guarantee global optimality.

These conditions may be stated as,

$$\left.\begin{aligned} &\frac{\partial f}{\partial x_j} - \sum_{i=1}^{m} \lambda_i \frac{\partial g_i}{\partial x_j} = 0 \qquad j = 1, 2, \ldots, n. \\ &g_i(x) - b_i \leqslant 0 \qquad i = 1, 2, \ldots, m, \end{aligned}\right. \tag{9.19}$$

$$\lambda_i \geqslant 0, \qquad i = 1, 2, \ldots, m, \tag{9.20}$$

and

$$\sum_{i=1}^{m} \lambda_i (g_i(x) - b_i) = 0 \tag{9.21}$$

The conditions (9.19), (9.20), (9.21) taken together define complementarity condition connecting the slacks and the Kuhn–Tucker multipliers associated with the inequality constraints. To exclude some pathological cases the above conditions also call for "Constraint Qualification". Constraint qualification is intimately connected with the deduction of Kuhn–Tucker conditions. In general terms this stipulates that there must exist an epsilon neighbourhood of such an optimum where for a point x, $g_i(x)$ are strictly less than b_i for all i and $g_i(x)$ are non-linear functions.

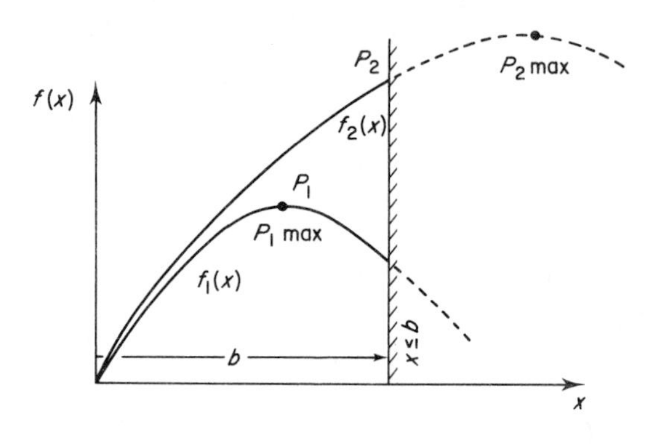

FIG. 9.1.

Kuhn–Tucker conditions involving the complementarity property may be illustrated by one-dimensional constrained optimization problems. Consider the function $f(x)$ shown in Fig. 9.1 and the problem

$$\begin{gathered} \text{Maximize } f(x) \\ \text{subject to } x \leqslant b. \end{gathered} \tag{9.22}$$

Assume for the moment that $f(x)$ is concave i.e. admits a global maximum. The unconstrained maximum of the function can occur within the region (closed half space) $x \leqslant b$ as shown, for the case where $f(x) = f_1(x)$: the maximum is then at the point P_1 max. On the other hand the unconstrained maximum for the case $f(x) = f_2(x)$ may lie outside this region; let this be P_2 max. Let x', x'' denote the x values at which the unconstrained maximum of $f_1(x)$ and $f_2(x)$ are taken and let P_1 and P_2 be the maxima of the two constrained maximization problems derived out of (9.22) and stated below.

$$\begin{aligned}&\text{Maximize } f_1(x)\\&\text{subject to } x \leqslant b,\end{aligned} \tag{9.23}$$

and

$$\begin{aligned}&\text{Maximize } f_2(x)\\&\text{subject to } x \leqslant b.\end{aligned} \tag{9.24}$$

The optimality relationship for these two cases may be summarized as follows.

(i) $P_1 = P_1$ max i.e. the unconstrained maximum of $f_1(x)$ at x' is also the maximum of the constrained problem stated in (9.23). In mathematical terms this implies that at P_1, $\partial f/\partial x = 0$ and $x \leqslant b$.

(ii) $P_2 \neq P_2$ max i.e. the unconstrained maximum at x'' is outside $x \leqslant b$ and the maximum is taken at P_2 on the surface $x = b$ which is a boundary point. In mathematical terms this implies that $\partial f/\partial x > 0$ (i.e. a rising slope) and $x = b$ at the point P_2.

This may be summed up by the following conditions

$$\text{either } \frac{\partial f}{\partial x} = 0 \text{ and } x \leqslant b \tag{9.25}$$

$$\text{or } \quad \frac{\partial f}{\partial x} > 0 \text{ and } x = b. \tag{9.26}$$

The Kuhn–Tucker conditions express this succinctly by the mathematical relationships,

$$\lambda(x - b) = 0, \tag{9.27}$$

where

$$\frac{\partial f}{\partial x} = \lambda, \lambda \geqslant 0 \quad \text{and} \quad x \leqslant b. \tag{9.28}$$

This is of course the complementarity relationship for the one dimensional case.

9.5 AN EXAMPLE ILLUSTRATING THE USE OF LAGRANGE AND KUHN–TUCKER MULTIPLIERS

A simple problem of magnetism is illustrated in Fig. 9.2. The Pole of a magnet is located at the co-ordinate point (4, 3) on a horizontal surface and the equipotential lines in the (x_1, x_2) plane are defined by the concentric circles

$$\phi(x_1, x_2) = (x_1 - 4)^2 + (x_2 - 3)^2. \tag{9.29}$$

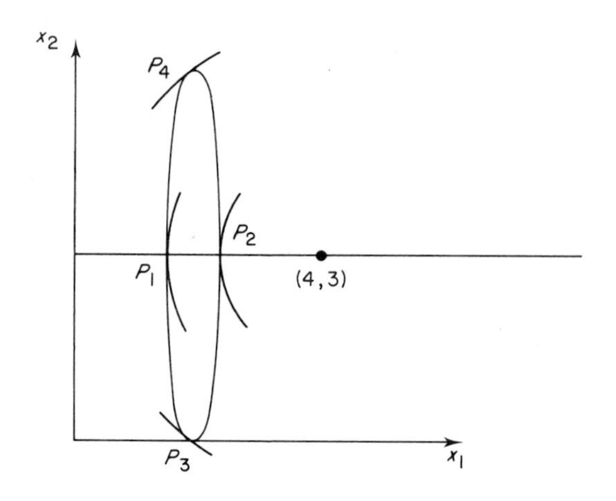

FIG. 9.2.

Let a steel ball be

(a) free to move in an elliptic path (groove) on this x_1, x_2 plane defined by the equation

$$g(x_1, x_2) = 36(x_1 - 2)^2 + (x_2 - 3)^2 = 9, \tag{9.30}$$

or (b) free to move in an elliptic area within the region

$$g(x_1, x_2) = 36(x_1 - 2)^2 + (x_2 - 3)^2 \leqslant 9. \tag{9.31}$$

Then the problem of minimizing $\phi(x)$ in (9.29) subject to (9.30) is a Lagrangean problem and that of minimizing $f(x)$ in (9.29) subject to (9.31) is a Kuhn–Tucker inequality problem. In both the cases the solution of the physical problem of determining the position of rest of the steel ball is sought. In both the cases the Lagrangean function to be investigated is

$$F(x_1, x_2, \lambda) = -\phi(x_1, x_2) - \lambda(g(x_1, x_2) - b). \tag{9.32}$$

To solve the first problem the following set of three non-linear equations in three variables need to be solved.

$$\frac{\partial F}{\partial x_1} = -[2(x_1 - 4) + \lambda 72(x_1 - 2)] = 0 \tag{9.33}$$

$$\frac{\partial F}{\partial x_2} = -[2(x_2 - 3) + 2.\lambda(x_2 - 3)] = 0, \tag{9.34}$$

$$\frac{\partial F}{\partial \lambda} = -[36(x_1 - 2)^2 + (x_2 - 3)^2 - 9] = 0. \tag{9.35}$$

From (9.34) it follows that if $x_2 = 3$ then λ and x_1 must be calculated from the remaining two equations (9.33), (9.35).

Setting $x_2 = 3$ in (9.35) $x_1 = \frac{3}{2}$ or $\frac{5}{2}$ and

$$\lambda = -\frac{1}{36}\left(\frac{x_1 - 4}{x_1 - 2}\right) = -\frac{5}{36} \quad \text{or} \quad +\frac{1}{12}.$$

These correspond to the points

$$P_1: \ x_1 = \tfrac{3}{2}, \ x_2 = 3, \ \lambda = -\tfrac{5}{36}$$

and

$$P_2: \ x_1 = \tfrac{5}{2}, \ x_2 = 3, \ \lambda - \tfrac{1}{12}.$$

If $x_2 \neq 3$ it follows from (9.34) that $\lambda = -1$. Substituting $\lambda = -1$ in (9.33) and (9.35) it follows that the two points

$$P_3: \quad x_1 = 1{\cdot}956, \quad x_2 = 0{\cdot}002, \quad \lambda = -1,$$

and

$$P_4: \quad x_1 = 1{\cdot}956, \quad x_2 = 5{\cdot}998, \quad \lambda = -1.$$

are obtained.

Of these four stationary points of the Lagrangean, the first two P_1, P_2 are the local optimum of the problem (a) and P_3, P_4 are points of inflexion or points of unstable rest.

In the problem (b), which is the Kuhn–Tucker problem, only the point P_2 satisfies the Kuhn–Tucker condition i.e.

$$\lambda = \frac{\partial f}{\partial x_1} = \frac{1}{12} > 0$$

and $g(x) = b$, i.e. $\lambda.(g(x) - b) = 0$. Hence this is an optimum point. The functions $\phi(x)$ and $g(x)$ are convex hence the local optimum in this case is also the global optimum.

EXERCISES

9.1 Let $x'Px$, $x'Qx$ be two quadratic forms involving n x-variables and let s, t, r be n-vectors. Then deduce the condition of optimality of the problems.

(i) Minimize $x'Qx + sx$
subject to $x'Px + tx = r$.

(ii) Minimize $x'Qx + sx$
subject to $x'Px + tx \leqslant r$
and $x \geqslant 0$.

9.2 Apply the Lagrange multiplier method to solve the problems

(i) Maximize $2x_1 + 3x_2$
subject to $2x_1^2 + 6x_2^2 = 12$

(ii) Maximize $x_1 + x_2$
subject to $-x_1^2 + x_2^2 - x_1 - x_2 = 0$.

CHAPTER 10

Convex Quadratic Programming: Its Application and Its Solution by the Use of Kuhn-Tucker Theory

KEYWORDS: Quadratic Form, Positive Definiteness, Kuhn–Tucker Conditions, Complementary Bases, Standard Tableau, Non-standard Tableau, Gaussian Algorithm, Diagonalization of Quadratic Form.

10.1 APPLICATIONS OF QUADRATIC PROGRAMMING

These have been discussed by Dantzig [15] and Wolfe [55]. The problems,

(a) regression analysis with non-negativity constraints and or constraints of upper or lower bounds on the regression parameters,

(b) to find the minimum of a general convex function subject to linear inequalities, where the convex function is twice differentiable and may be locally approximated by a positive definite quadratic form,

(c) production planning by maximizing profit when the marginal cost is linearly varying, and also the production function is linearly varying with the problem variable,

(d) in a linear model, for given ranges of expected profits, to find the solution which minimises the variance (risk level) of such profits,

all admit formulation in the format of convex quadratic programming. Quadratic programming has been also applied in such diverse problem areas as optimal use of milk in Netherlands [34], or optimal alignment of the vertical profile of highways [42].

10.2 STATEMENT OF THE QUADRATIC PROGRAMMING PROBLEM

The quadratic programming problem may be stated as

$$\text{Find the maximum of } f(x) = \sum_{j=1}^{n} p_j x_j - \tfrac{1}{2} \sum_{i=1}^{n} \sum_{j=1}^{n} q_{ij} x_i x_j \qquad (10.1)$$

$$\text{subject to the constraints,} \quad \sum_{j=1}^{n} a_{ij}x_j \leqslant b_i \qquad i = 1, 2, \ldots, m, \tag{10.2}$$

and $x_1, x_2, \ldots, x_n \geqslant 0$.

In matrix notation this can be presented as

$$\text{Maximize } f(x) = p'x - \tfrac{1}{2}x'Qx \tag{10.3}$$

$$\text{subject to } Ax \leqslant b, x \geqslant 0, \tag{10.4}$$

where p, b are n and m vectors, A is an $m \times n$ matrix Q an $n \times n$ matrix of quadratic form expressed with an x-vector of n unknown variables. In this chapter only the special case where Q is positive semi-definite (hence the corresponding problem admits global optimum) is considered. This is of course the most frequently occurring case. A method of testing this definiteness property of the quadratic form associated with a square matrix say Q is discussed later in this chapter in Section 10.6.

10.3 REPRESENTATION OF THE PROBLEM AND ITS OPTIMALITY CONDITIONS

Introducing Kuhn–Tucker multipliers v a vector of m unknowns for the m inequality constraints and u a vector of n unknowns for the non-negativity constraints on the x variables, viz. $-x \leqslant 0$ the following Lagrangean may be set up.

$$F(x, v, u) = p'x - \tfrac{1}{2}x'Qx - v'(Ax - b) + u'x. \tag{10.5}$$

Introduce a slack vector y of m non-negative components corresponding to m inequalities, the optimality relationship for the problem by the application of Kuhn–Tucker conditions may be stated as,

$$p - Qx - v'A + u = 0 \tag{10.6}$$

$$Ax + Iy = b \tag{10.7}$$

$$v'(Ax - b) = v'y = 0 \tag{10.8}$$

$$u'x = 0 \tag{10.9}$$

$$u, v, x, y \geqslant 0. \tag{10.10}$$

To represent the full problem as a set of linear equations expressed in the tableau form, a linear expression may be derived for the objective function. If the complementarity conditions as stated in (10.8), (10.9), (10.10) are satisfied, then introducing the variable u_0 it follows

$$
\begin{aligned}
u_0 &= 2f(x) = 2p'x - x'Qx = p'x + x'(-u + Qx + v'A) - x'Qx \\
&= p'x - x'u + x'A'v = p'x - x'u + v'Ax \\
&= p'x - x'u + v'(b - y) = p'x + v'b - (x'u + v'y) \\
&= p'x + v'b \qquad (10.11)
\end{aligned}
$$

when (10.8), (10.9), (10.10) hold.

The relations (10.11), (10.6), (10.7) may be rewritten as

$$u_o = 2f(x) = \sum_{j=1}^{n} p_j x_j + \sum_{k=1}^{m} v_k b_k \qquad (10.12)$$

$$u_i = p_i + \sum_{j=1}^{n} q_{ij} x_j + \sum_{k=1}^{m} a_{ki} v_k \qquad i = 1, 2, \ldots, n \qquad (10.13)$$

$$y_i = b_i - \sum_{j=1}^{n} a_{ij} x_j, \qquad i = 1, 2, \ldots, m. \qquad (10.14)$$

and are displayed in the following Tableau, 10.00.

	$-x_0$	$-x_1$	$-x_2$	$-x_n$	$-v_1$	$-v_2$	$-v_m$
$2f(x) = u_0$	0	$-p_1$	$-p_2$	$-p_n$	$-b_1$	$-b_2$	$-b_m$
u_1	$-p_1$	$-q_{11}$	$-q_{12}$	$-q_{1n}$	$-a_{11}$	$-a_{21}$	$-a_{m1}$
u_2	$-p_2$	$-q_{21}$	$-q_{22}$	$-q_{2n}$	$-a_{12}$	$-a_{22}$	$-a_{m2}$
		$-Q$				$-A'$	
u_n	$-p_n$	$-q_{n1}$	$-q_{n2}$	$-q_{nn}$	$-a_{1n}$	$-a_{2n}$	$-a_{mn}$
y_1	b_1	a_{11}	a_{12}	a_{1n}			
y_2	b_2	a_{21}	a_{22}	a_{2n}			
		A				0	
y_m	b_m	a_{m1}	a_{m2}	a_{mn}			

TABLEAU 10.0

It should be noted that the above tableau possesses a symmetry and a skew symmetry property. The Kuhn–Tucker condition states that of all the possible vertices of the simplex defined by this tableau, the vertex in which the complementarity condition holds and the solution values for x, y the primal and u, v the dual variables are non-negative, yields the optimal solution which is contained in the corresponding tableau.

10.4 AN ALGORITHM FOR SOLVING QUADRATIC PROGRAMMING PROBLEM [51]

Except for Beale's method of quadratic programming which does not invoke Kuhn–Tucker's optimality criterion, all the other well known methods of quadratic programming solve the problem by

(i) trying to satisfy the conditions (10.6) through (10.10),
(ii) and improving the function value in successive iterations (primal methods).

In this section an algorithm due to Van-de-Panne and Whinston [51] is described: the description provides the pivot selection rules and subsequent pivotal transformations leading to the optimal quadratic program.

To start with, the following need to be defined.

1. Standard tableau

If for each x or y (i.e. primal variable) in the basis its corresponding u or v (dual variable) is non-basic and vice-versa, then such a tableau possesses a symmetry/skew-symmetry property and is called a "Standard Tableau".

2. Non-standard tableau

A tableau in which one pair of x, u or y, v variable together appear in the basis is called a "Non-standard Tableau" (or a "Nearly Complementary Tableau").

The pivot choice rules for transforming the tableau may now be stated as follows.

Rule 1. Choice of variable to enter the basis.

In the case that the tableau is in standard form, select as the new basic variable the y or x-variable having the largest negative corresponding basic u or v variable solution value. In the case that the tableau is not in standard form introduce as the new basic variable the u or v variable of the non-basic pair from the list of the non-basic variables.

Rule 2. Choice of the variable to leave the basis.

Select as the variable to leave the basis, from the set of basic x, y variables and u_s or v_s the variable which first becomes zero upon introducing the new basic variable by a positive amount. Here u_s or v_s is the dual variable corresponding to the x or y variable in the standard tableau, and in a non-standard tableau it is the dual variable of the basic pair.

If I denotes the set of indices of the rows in which the x or y variables are pivoted and the variable $u_s(v_s)$ is pivoted, then the pivot choice rule may be expressed as:

Choose as the pivot row r such that,

$$\min_{i \in I} \left\{ \frac{\alpha_{i0}}{\alpha_{ij}} \,\middle|\, \frac{\alpha_{i0}}{\alpha_{ij}} \geqslant 0, \quad \text{and} \quad \alpha_{ij} \neq 0 \right\} = \frac{\alpha_{r0}}{\alpha_{rj}}, \tag{10.15}$$

where α_{ij} denotes the (i, j)th element of any intermediate tableau and j is the column corresponding to the variable chosen to enter the basis.

Having chosen the pivot element of course the tableau is transformed by transformation rules identical to those used in the simplex method of linear programming.

10.5 A WORKED EXAMPLE

Consider the problem,

$$\begin{aligned} &\text{Maximize } f(x) = 6x_1 - 2x_1^2 + 2x_1x_2 - 2x_2^2 \\ &\text{subject to} \qquad x_1 + x_2 \leqslant 2 \\ &\text{and} \qquad\qquad x_1, x_2 \geqslant 0. \end{aligned}$$

This may be expressed in vector notation as

$$f(x) = \underset{p'}{[6, 0]} \underset{x}{\begin{bmatrix} x_1 \\ x_2 \end{bmatrix}} - \tfrac{1}{2} \underset{x'}{[x_1, x_2]} \underset{Q}{\begin{bmatrix} 4 & -2 \\ -2 & 4 \end{bmatrix}} \underset{x}{\begin{bmatrix} x_1 \\ x_2 \end{bmatrix}}$$

$$\underset{A}{[1, 1]} \underset{x}{\begin{bmatrix} x_1 \\ x_2 \end{bmatrix}} \leqslant \underset{b}{2} \qquad x_1, x_2 \geqslant 0.$$

The problem is set up in Tableau 10.1 which is a standard tableau and the accompanying diagram illustrates the unconstrained and constrained maximum of $f(x)$ which defines a system of elipses.

At the point $(0, 0)$ Tableau 10.1 the value of $u_1 = -6$ and hence x_1 is chosen to come into the basis. The rationale for such a step may be expressed from the first principles as follows. Setting all the other variables to zero

$$f(x) = 6x_1 - \tfrac{1}{2}.4.(x_1)^2 \quad \text{and} \quad \frac{\partial f}{\partial x_1} = 6 - 4x_1 \tag{10.16}$$

hence the function value increases with x_1.

	$-x_0$	$-x_1\downarrow$	$-x_2$	$-v_1$
u_0	0	-6	0	-2
u_1	-6	$\underline{\underline{-4}}$	2	-1
u_2	0	2	-4	-1
y_1	2	1	1	0

STANDARD TABLEAU 10.1

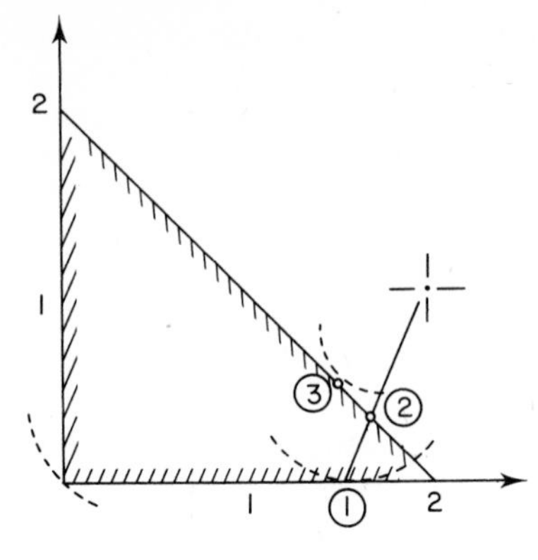

FIG. 10.1

From $y_1 = 2 - x_1 - x_2$ it follows that x_1 can go up to $\frac{2}{1} = 2$ and from $\partial f/\partial x_1 = 0 = 6 - 4x_1$ the maximum of $f(x)$ is obtained for $x_1 = \frac{6}{4} = \frac{3}{2}$ (look at row 2). Therefore the min of the ratios $\{\frac{3}{2}, \frac{2}{1}\}$ is taken and the element -4 in Tableau 10.1 is chosen for pivoting and Tableau 10.2 is obtained after transformation. After a sequence of pivot choices and transformations the optimum solution is obtained and this is contained in Tableau 10.4.

	$-x_0$	$-u_1$	$-x_2\downarrow$	$-v_1$
u_0	9	$-\frac{3}{2}$	-3	$-\frac{1}{2}$
x_1	$\frac{3}{2}$	$-\frac{1}{4}$	$-\frac{1}{2}$	$\frac{1}{4}$
u_2	-3	$\frac{1}{2}$	-3	$-\frac{3}{2}$
y_1	$\frac{1}{2}$	$\frac{1}{4}$	$\frac{3}{2}$	$-\frac{1}{4}$

STANDARD TABLEAU 10.2

	$-x_0$	$-u_1$	$-y_1$	$-v_1\downarrow$
u_0	10	-1	2	-1
x_1	$\frac{5}{3}$	$-\frac{1}{6}$	$\frac{1}{3}$	$\frac{1}{6}$
u_2	-2	1	2	-2
x_2	$\frac{1}{3}$	$\frac{1}{6}$	$\frac{2}{3}$	$-\frac{1}{6}$

NON-STANDARD TABLEAU 10.3

True function value in Tableau 10.3 is $2f(x) = 10 - u_2 x_2 = 10\frac{2}{3}$

	$-x_0$	$-u_1$	$-y_1$	$-u_2$
u_0	11	$-\frac{3}{2}$	1	$-\frac{1}{2}$
x_1	$\frac{3}{2}$	$-\frac{1}{12}$	$\frac{1}{2}$	$\frac{1}{12}$
v_1	1	$-\frac{1}{2}$	-1	$-\frac{1}{2}$
x_2	$\frac{1}{2}$	$\frac{1}{12}$	$\frac{1}{2}$	$-\frac{1}{12}$

Optimal

STANDARD TABLEAU 10.4

10.6 POSITIVE DEFINITENESS OF QUADRATIC FORMS [46]

A quadratic form $x'Qx$ associated with the symmetric matrix Q is said to be positive definite or positive semi-definite if the inequality (10.17)

$$x'Qx \geqslant 0 \tag{10.17}$$

holds strictly or otherwise, for all x in E^n.

Two tests may be suggested for this,

(a) A sufficient condition for positive definiteness is given by the Hadamard–Greschgorin condition for diagonal dominance, viz. if

$$q_{ii} \geqslant \sum_{j \neq i} |q_{ij}| \qquad \text{for all } i = 1, 2, \ldots, n, \tag{10.18}$$

then all the diagonal terms are non-negative and dominant.

(b) The "Gaussian Algorithm" may be applied to express a quadratic form as the sum of squares and hence establish if the form is definite or not.

For an illustration of the procedure consider the form

$$F_3 = x'Qx = [x_1, x_2, x_3] \times \begin{bmatrix} q_{11} & & \\ q_{21} & q_{22} & \\ q_{31} & q_{32} & q_{33} \end{bmatrix} \times \begin{bmatrix} x_1 \\ x_2 \\ x_3 \end{bmatrix} \tag{10.19}$$

$$\begin{aligned} &= q_{11}x_1^2 \\ &\quad + 2q_{21}x_1x_2 + q_{22}x_2^2 \\ &\quad + 2q_{31}x_1x_3 + 2q_{32}x_3x_2 + q_{33}x_3^2. \end{aligned} \tag{10.20}$$

For $q_{11} \neq 0$ this may be expressed as

$$\begin{aligned} F_3 = \frac{1}{q_{11}} &\{q_{11}^2x_1^2 + 2q_{21}q_{11}x_1x_2 + 2q_{31}q_{11}x_3x_1 + \\ &+ 2q_{31}q_{21}x_3x_2 + q_{21}^2x_2^2 + q_{31}^2x_3^2\} \\ &+ \left(q_{22} - \frac{q_{21}^2}{q_{11}}\right)x_2^2 \\ &+ 2\left(q_{32} - \frac{q_{31}q_{21}}{q_{11}}\right)x_3x_2 + \left(q_{33} - \frac{q_{31}^2}{q_{11}}\right)x_3^2, \end{aligned} \tag{10.21}$$

$$F_3 = \frac{1}{q_{11}}\{q_{11}x_1 + q_{21}x_2 + q_{31}x_3\}^2 + (x_2, x_3) \times \begin{bmatrix} p_{22} & \\ p_{32} & p_{33} \end{bmatrix} \times \begin{bmatrix} x_2 \\ x_3 \end{bmatrix}, \text{ say}$$

i.e.,

$$F_3 = \frac{1}{q_{11}} \times \{\text{square of a linear form}\} + F'_2 \tag{10.22}$$

where

$$F'_2 = (x_2, x_3) \times \begin{bmatrix} p_{22} & \\ p_{32} & p_{33} \end{bmatrix} \times \begin{bmatrix} x_2 \\ x_3 \end{bmatrix}, \quad \text{and} \quad p_{22} = \left(q_{22} - \frac{q_{21}^2}{q_{11}}\right) \text{ etc.}$$

From (10.22) it follows that F_3 is non-negative for all x if F'_2 is non-negative in all the remaining x and the first term is always non-negative. This is true if

(*i*) $q_{11} \geqslant 0$

and

(ii) $F'_2 \geqslant 0$.

Thus the test for positive definiteness becomes that of finding a positive pivot q_{11} and applying a Gaussian algorithm and reducing the form to square of a linear form and a reduced quadratic form which must again be positive (semi) definite. If at any stage of carrying out this process no positive pivot can be found then the form is not positive definite. If there are some zeros and some positive pivots then it is semi definite, and if there remain any negative pivots then it is indefinite. At the end of applying the algorithm to the entire Q matrix the resulting expression reduces the original quadratic form to the sum of squares of some new linear forms. The Gaussian algorithm is now illustrated by the following example. Given the Q matrix

$$Q = \begin{bmatrix} 5 & & \\ 2 & 1 & \\ 10 & 3 & 21 \\ & & [38] \end{bmatrix} \tag{10.23}$$

(we assume $q_{33} = 21$, or 38 leading to two different Q matrices). We wish to diagonalize it and establish its definiteness.

$$F_3 = [x_1, x_2, x_3] \times \begin{bmatrix} 5 & & \\ 2 & 1 & \\ 10 & 3 & 21 \\ & & \text{or } [38] \end{bmatrix} \times \begin{bmatrix} x_1 \\ x_2 \\ x_3 \end{bmatrix},$$

or

$$\begin{aligned} F_3 = \tfrac{1}{5}\{5x_1 + 2x_2 + 10x_3\}^2 \\ + \tfrac{1}{5}x_2^2 \\ - 2x_2x_3 + x_3^2 \end{aligned}$$

or $[18x_3^2]$, or

$$\begin{aligned} F_3 = \tfrac{1}{5}\{5x_1 + 2x^2 + 10x_3\}^2 \\ + 5\{\tfrac{1}{5}x_2 - x_3\}^2 \\ - 4x_3^2 \\ \text{or } [13x_3^2]. \end{aligned}$$

Hence in the first case the form is indefinite and in the second case the form is positive definite.

EXERCISES

10.1 Solve the quadratic programming problem

$$\begin{aligned} &\text{Minimize } Q(x) = \tfrac{1}{2}x_1^2 + \tfrac{1}{2}x_2^2 - x_1 - 2x_2 \\ &\text{subject to} \quad 2x_1 + 3x_2 \leqslant 6 \\ &\qquad\qquad\quad x_1 + 4x_2 \leqslant 5 \\ &\text{and} \qquad\quad\ x_1, x_2 \geqslant 0. \end{aligned}$$

Illustrate your solution graphically and comment on the problem of maximizing $Q(x)$ subject to the same constraints.

10.2 Comment on the possible methods of solution of the problem

$$\begin{aligned} &\text{Minimize } 2x_1x_2 + 3x_2x_3 + 4x_3x_4 \\ &\text{subject to } x_1 + x_2 + 2x_3 + 3x_4 \leqslant 5 \\ &\text{and} \qquad x_1, x_2, x_3, x_4 \geqslant 0. \end{aligned}$$

10.3 Express the quadratic form

$$Q(x) = 2x_1^2 + 5x_2^2 + 6x_3^2 + 6x_1x_2 + 6x_2x_3 + 6x_3x_1,$$

as a sum of squares and hence comment on the problem of finding the minimum of $\bar{Q}(x)$ subject to $x_1, x_2, x_3 \geqslant 0$, where

$$\bar{Q}(x) = Q(x) + x_1 + x_2 + x_3.$$

CHAPTER 11

Linear Programming, Quadratic Programming, Theory of Games, and The Fundamental Problem: Algebra and Combinatorics of Pivot Theory for such Problems

KEYWORDS: Matrix Game, Pay-off Table, Pure Strategy, Mixed Strategy, Two Person Games, Zero-sum Games, Non-zero Sum Games, Bimatrix-Games, Min-max Theorem, Equilibrium Point, Fundamental Problem, Nearly Complementary Bases.

11.1 AN APPROACH TO UNIFY THE PROGRAMMING AND GAMES PROBLEMS AND THEIR SOLUTIONS

The works of Tucker [17], Lemke [33], Dantzig and Cottle [13], in the last decade have led to a unifying approach towards tackling such problems as Linear Programming, Quadratic Programming, Games Theory and a few others [20]. The attempts to put these problems in one format: the Fundamental Problem (see Section 11.6) and the development of the Complementary Pivot Theory for the solution of this problem is by no means just an exercise in abstraction. On the contrary it provides considerable insight into the theory and design of algorithms for the solution of these problems. That such theory finds use in devising new solution techniques out of the existing procedures is strikingly illustrated by Wolfe's approach to Quadratic Programming [55] whereby an existing Rand Code to solve Linear Programs was modified by a few instructions to solve quadratic programming problems.

The topic of Games Theory has not been considered elsewhere in the book except for passing comments in the Introduction. In the next four sections some aspects of the problem of Matrix Games have been discussed; we concern ourselves primarily with the mathematics underlying the games and how the computational schemes are connected with two other problems of this genre, viz., Linear Programming and Quadratic Programming. In Section 11.5 the three central problems, viz., Linear Programming, Quadratic Programming, and Matrix Games are stated in the form of the Fundamental Problem. Finally, two schemes for solving the Fundamental Problem are presented in Section 11.6.

11.2 MATRIX GAMES, THEIR APPLICATIONS, THEORY AND SOME TERMINOLOGY

The mathematical formulation of competitive situations were considered independently by J. von Neumann [53], and Emil Borel in the 1920s. However, the min–max theorem which states a fundamental property of games, the solution technique of matrix games, as well as the equivalence of linear programs and matrix games were rigorously developed by J. von Neumann. Hence all the text books on this topic start by acknowledging credit where it is due.

The work of J. von Neumann and Oskar Morgenstern [53] had raised the hopes of a number of people working in such areas of social sciences as psychology, economics, operations research etc., that this powerful mathematical tool for analysing problems would find increasing use. Indeed it was conjectured that there would be an explosion in the application of games theory. The facts have unfortunately proven to be otherwise. We may quote at least one of the leading practitioners of OR: "In practising Operations Research, we have found Games Theory does not contribute any *managerial insights* to real competitive and co-operative decision making behaviour" (Harvey M. Wagner [54]).

The book by Luce and Raiffa [35] is perhaps the most authorative work in this field after that of J. von Neumann and Morgenstern. In this the authors are often at pains to point out the existence of such imponderables as "rational players". They also consider the motives of competing players, mathematically advisable strategies of (correct?) players and the associated anomalies in the context of the human behaviour. The mathematics underlying games theory is in comparison much easier to understand. For instance the concept of mixed strategy, whereby based on the outcome of some hypothetical experiment a player chooses some of the alternatives, is simple to formulate mathematically but it is difficult to comprehend in practice. In the present chapter we restrict ourselves only to the computational aspects of the problem of matrix games. Some important aspects of games theory, viz., description of a complex game in the extensive form, reducing a game into a normal form whereby a complex game is reduced by enumeration to a matrix game, games with incomplete information and infinite games have not been discussed, for an extensive treatment of these topics the reader should refer to [35].

Given a matrix A of m rows and n columns, and having elements representing such things as money or men or any item in which gain or loss may be quantified, and given two hypothetical persons (players) Row Player (RP) and Column Player (CP), it may be postulated that RP has the alternative of choosing an integer row index i $(1 \leqslant i \leqslant m)$ and CP has the alternatives of choosing an integer column index $j(1 \leqslant j \leqslant n)$. After RP, CP have chosen their respective (i, j) indices a payment of a_{ij} is made by CP to RP and if the quantity a_{ij} is negative then the payment is made by RP to CP. The process of choosing (i, j) and paying a_{ij} is called a "move" of the game. Each person's individual choice is called a "Strategy" and the table of a_{ij} making up the matrix A is called the "pay-off table"; the game so defined is a "matrix game", and it is further a "two-person game" (played by RP–CP) and a "Zero-Sun Game" as RP's gain is CP's loss and vice versa.

Consider the following pay-off table, 11.1 in which RP and CP each has

TABLE 11.1. Pay-off table

RP ↓ / CP →	$j = 1$	$j = 2$	$j = 3$
$i = 1$	30	0	−10
$i = 2$	20	10	20
$i = 3$	−10	0	30

three choices ($m = n = 3$). Assume that each of the players wishes to maximise their respective gains. Further each player knows of similar motive in his adversary and hence if he finds that he must lose he wishes to at least minimise his loss. In this example it is obvious that RP wins and CP loses. The minimum of the column maxima $a_{22} = 10$, $i = 2$, $j = 2$ is the same as the maximum of the row minima. This implies that the game has a saddle point (equilibrium point) for this "pure strategy". The arguments for each player to choose their respective strategies are as follows:

RP wishes to maximise the minimum quantity he is going to gain, $i = 2$ assures him a gain of at least 10 whereas for $i = 1$, or 3 he might lose -10 if CP plays $j = 3$, or 1. On his part CP wishes to minimize the maximum loss he might incur. Thus for $j = 2$ he is assured that he does not lose more than $a_{22} = 10$. Another way to specify the strategy of each player is to introduce choice variables $x_{n+1}, x_{n+2}, x_{n+3}$ and y_1, y_2, y_3 for RP and CP and to introduce the two scalars g, h where, h denotes the amount RP is sure to win and g

denotes the amount CP is bound to lose, such that,

$$
\begin{aligned}
30y_1 + 0y_2 - 10y_3 &\leqslant g \\
20y_1 + 10y_2 + 20y_3 &\leqslant g \\
-10y_1 + 0y_2 + 30y_3 &\leqslant g
\end{aligned}
$$

and

$$
\begin{aligned}
30x_{n+1} + 20x_{n+2} - 10x_{n+3} &\geqslant h \\
0x_{n+1} + 10x_{n+2} + 0x_{n+3} &\geqslant h \\
-10x_{n+1} + 20x_{n+2} + 30x_{n+3} &\geqslant h
\end{aligned}
\tag{11.1}
$$

further

$$y_1 + y_2 + y_3 = 1, \qquad y_1, y_2, y_3 \geqslant 0$$

and

$$x_{n+1} + x_{n+2} + x_{n+3} = 1, \qquad x_{n+1}, x_{n+2}, x_{n+3} \geqslant 0.$$

In this example it so happens that one x_{n+i} and one y_j may be chosen and be made equal to 1 ($x_{n+2} = 1$, $y_2 = 1$) and the rest of the x, y set to zero; then $g = h = 10$ is a solution of the game. The last quantity is known as the "value" of the game. In this case RP and CP are said to have a pure strategy (unique choice) each, and the game is determined. However, it may be that the decision variables x_{n+i}, y_j may not necessarily take integer 0–1 values but take values in the range $0 \leqslant x_{n+i}, y_j \leqslant 1$. In this case these variables may be interpreted as representing the probabilities with which the *i*th or the *j*th alternatives are chosen. The quantity $\Sigma a_{ij}y_j$ then represents the "Mathematical Expectation" of the row player RP for a given i and various possible choices of strategies (j) for the column player CP. Similarly $\Sigma_i a_{ij}x_{n+i}$ is the mathematical expectation of CP. The vectors x, y of the decision variables in this case are said to define "Mixed Strategies" for RP and CP respectively.

The min–max theorem concerning Matrix Games and some of the properties which follow directly as a consequence of this may now be stated.

Given the row and column player's problems,

$$
\begin{aligned}
\text{RP:}\quad & a_{11}x_{n+1} + a_{21}x_{n+2} + \ldots + a_{m1}x_{n+m} \leqslant h \\
& a_{12}x_{n+1} + a_{22}x_{n+2} + \ldots + a_{m2}x_{n+m} \geqslant h \\
& \quad \vdots \qquad\qquad \ldots \qquad\qquad \vdots \\
& a_{1n}x_{n+1} + a_{2n}x_{n+2} + \ldots + a_{mn}x_{n+m} \geqslant h \\
& x_{n+1} + \quad x_{n+2} + \ldots + x_{n+m} \quad = 1 \\
\text{and}\quad & x_{n+1}, x_{n+2}, \ldots, x_{n+m} \geqslant 0,
\end{aligned}
\tag{11.2}
$$

$$
\begin{aligned}
\text{CP:}\quad & a_{11}y_1 + a_{12}y_2 + \ldots + a_{1n}y_n \leqslant g \\
& a_{21}y_1 + a_{22}y_2 + \ldots + a_{2n}y_n \leqslant g \\
& \quad\cdot \qquad\qquad \cdots \qquad \cdot \\
& \quad\cdot \qquad\qquad\qquad\quad \cdot \\
& a_{m1}y_1 + a_{m2}y_2 + \ldots + a_{mn}y_n \leqslant g \\
& \quad y_1 + \quad y_2 + \ldots + \quad y_n = 1 \\
\text{and}\quad & y_1, y_2, \ldots, y_n \geqslant 0,
\end{aligned}
\tag{11.3}
$$

if each player plays conservatively then the largest floor for RP i.e., maximum of h is the smallest ceiling for CP, i.e., minimum of g. Further, the mixed strategy for each player solves a linear program and its dual. When such optimal strategies have been chosen the following relationship holds,

$$
w = \max_{x} \min_{y|x} \sum_i \sum_j a_{ij}x_{n+i}y_j = \min_{y} \max_{x|y} \sum_i \sum_j a_{ij}x_{n+i}y_j. \tag{11.4}
$$

In (11.4) the notation $y|x$ should be read as "y given x"; the quantity w is known as the value of the game. An either/or relation immediately follows from this and may be stated as:

If CP's optimal strategy yields a strict inequality in the ith relation of (11.3) then RP's optimal strategy must have $x_{n+i} = 0$; if RP's optimal strategy has s strict inequality in the jth relation of (11.2) then the CP's optimal strategy must have $y_j = 0$. This is proved later on in this section.

Let the inequalities (11.2), (11.3) be rewritten as equalities with surplus $x_j (1 \leqslant j \leqslant n)$ and slack $y_{n+1} (n + 1 \leqslant n + i \leqslant n + m)$ variables,

$$
\begin{aligned}
-x_1 \qquad\qquad & + a_{11}x_{n+1} + a_{21}x_{n+2} \cdots + a_{m1}x_{n+m} = h \\
-x_2 \qquad & + a_{12}x_{n+1} + a_{22}x_{n+2} \cdots + a_{m2}x_{n+m} = h \\
-x_n & + a_{1n}x_{n+1} + a_{2n}x_{n+2} \cdots + a_{mn}x_{n+m} = h \\
& \quad x_{n+1} + \quad x_{n+2} \cdots + \quad x_{n+m} = 1 \\
& x_1, x_2, \ldots, x_n, x_{n+1}, \ldots, x_{n+m} \geqslant 0,
\end{aligned}
\tag{11.5}
$$

and

$$
\begin{aligned}
& a_{11}y_1 + a_{12}y_2 + \ldots + a_{1n}y_n + y_{n+1} && = g \\
& a_{21}y_1 + a_{22}y_2 + \ldots + a_{2n}y_n \qquad + y_{n+2} && = g \\
& \cdot \qquad \cdot \qquad \cdot \\
& \cdot \qquad \cdot \qquad \cdot \\
& \cdot \qquad \cdot \qquad \cdot \\
& a_{m1}y_1 + a_{m2}y_2 + \ldots + a_{mn}y_n \qquad\qquad + y_{m+n} && = g \\
& \quad y_1 + \quad y_2 + \ldots + \quad y_n && = 1 \\
& y_1, y^2, \ldots, y_n, y_{n+1}, \ldots, y_{m+n} \geqslant 0.
\end{aligned}
\tag{11.6}
$$

For any pair of solutions of (11.5) and (11.6) it follows,

$$
\begin{aligned}
&y_1(x_1) + y_2(x_2) \ldots + y_n(x_n) + g(-1) \\
&\quad = y_1(a_{11}x_{n+1} + a_{21}x_{n+2} + \ldots + a_{m1}x_{n+m} - h) \\
&\qquad + y_2(a_{12}x_{n+1} + a_{22}x_{n+2} + \ldots + a_{m2}x_{n+m} - h) \\
&\qquad \ldots \\
&\qquad + y_n(a_{1n}x_{n+1} + a_{2n}x_{n+2} + \ldots + a_{mn}x_{n+m} - h) \\
&\qquad + g(-x_{n+1} - x_{n+2} \cdots - x_{n+m}) \\
&\quad = x_{n+1}(a_{11}y_1 + a_{12}y_2 + \ldots, a_{1n}y_n - g) \\
&\qquad x_{n+2}(a_{21}y_1 + a_{22}y_2 + \ldots + a_{2n}y_n - g) \\
&\qquad \cdot \qquad \cdot \qquad \cdot \\
&\qquad \cdot \qquad \cdot \qquad \cdot \\
&\qquad \cdot \qquad \cdot \qquad \cdot \\
&\qquad x_{n+m}(a_{m1}y_1 + a_{m2}y_2 + \ldots + a_{mn}y_n - g) \\
&\qquad\quad + (\quad y_1 + \quad y_2 + \ldots + \quad y_n) \times (-h) \\
&\quad = x_{n+1}(-y_{n+1}) + x_{n+2}(-y_{n+2}) + \ldots + x_{n+m}(-y_{n+m}) + 1\,(-h).
\end{aligned}
$$

Hence,

$$(g - h) = y_1x_1 + y_2x_2 + \ldots + y_nx_n + y_{n+1}x_{n+1} + \ldots + y_{n+m}x_{n+m}. \tag{11.7}$$

The equation (11.7) is said to be the "key equation" for the two systems. For the optimal solution to the game we have $g = h$, $x \geqslant 0$ and $y \geqslant 0$; the complementary relationship stated earlier therefore follows from the key equation (11.7). Note that of all the feasible basic solutions to the game (there maybe a maximum of $(n + m)!/(n!m!)$ such basic solutions) only those which satisfy the complementary condition lead to one or more pairs of optimum mixed strategies. (As in Linear Programming because of degeneracy the possibility of multiple optima has to be admitted.)

As an example consider the problem specified by the pay-off matrix,

$$\begin{bmatrix} 3 & 4 & 8 \\ 6 & 5 & 4 \end{bmatrix}$$

which has not got an equilibrium point in terms of a pure strategy, viz.,

$$\max_i \{\min_j a_{ij}\} \neq \min_j \{\max_i a_{ij}\}. \tag{11.8}$$

However, the following pair of strategies for the RP and CP solves the problem in terms of mixed strategies,

$$\text{RP: } x_1 = 3/5, x_2 = 0, x_3 = 0, x_4 = 1/5, x_5 = 4/5$$

$$\text{CP: } y_1 = 0, y_2 = 4/5, y_3 = 1/5, y_4 = 0, y_5 = 0$$

This problem is again considered in the Section 11.4.

Games of cooperation

All the problems of Matrix Games do not necessarily relate to competing situations; there are certain games of co-operation where one person's gain is not equal to the other person's loss. For instance, consider the following bimatrix game defined for the two players RPA and CPB, and specified by the two matrices A and B each of dimension $m \times n$. The row player RPA has the option of choosing any one row of A $(1 \leqslant i \leqslant m)$ and the column player CPB has the option of choosing any one column of B $(1 \leqslant j \leqslant n)$. If each player uses their ith and jth respective pure strategies, then their losses are defined as a_{ij}, b_{ij} respectively. If, however, each player uses mixed strategies,

$$x : (x_1, x_2, \ldots x_m) \geqslant 0, \qquad \sum_{i=1}^{m} x_i = 1,$$

and

$$y : (y_1, y_2, \ldots y_n) \geqslant 0, \qquad \sum_{j=1}^{n} y_j = 1,$$

then their expected losses are $\Sigma_i\Sigma_j a_{ij}x_iy_j$ and $\Sigma_i\Sigma_j b_{ij}x_iy_j$ respectively. A pair of mixed strategies (x^0, y^0) for this problem is a Nash Equilibrium Point [35] if the following relationship holds,

$$\sum_i \sum_j a_{ij}x_i^0 y_j^0 \leqslant \sum_i \sum_j a_{ij}x_i y_j^0, \quad \text{for all mixed strategies } x, \tag{11.9}$$

and

$$\sum_i \sum_j b_{ij}x_i^0 y_j^0 \leqslant \sum_i \sum_j b_{ij}x_i^0 y_j \quad \text{for all mixed strategies } y. \tag{11.10}$$

As in the ordinary two-person zero-sum game, addition of a constant quantity to any of the two matrices does not change the equilibrium point. Thus the optimal mixed strategy (x^0, y^0) satisfying (11.9), (11.10) is also optimal for the problem with matrices A', B' such that

$$a'_{ij} = a_{ij} + \alpha \quad \text{for all} \quad i, j \tag{11.11}$$

$$b'_{ij} = b_{ij} + \beta$$

where α, β are constants. By solving the following set of equations and inequalities the bimatrix game may be solved, [33]

$$\begin{aligned} u &= Ay - e_m, \qquad u \geqslant 0, y \geqslant 0, \\ v &= B'x - e_n, \qquad v \geqslant 0, x \geqslant 0, \end{aligned} \tag{11.12}$$

and $xu + vy = 0$ where e_m, e_n are m and n dimensional vectors each component of which are unity.

In Section 11.5 this problem is put in the format of the Fundamental Problem.

11.3 TRANSFORMING A MATRIX GAME TO LINEAR PROGRAM AND VICE VERSA

The equivalence of matrix games and linear programs were discovered by Von-Neumann in the spring of 1947 and this was communicated to G. B. Dantzig; this is also the year when Dantzig presented his simplex method.

Given the two-person zero-sum game as specified by the pay-off matrix A, and the alternatives $x_{n+1} \ldots x_{n+m}$ for RP and $y_1 \ldots y_n$ for CP (see last section), the row player's problem is restated as,

$$\begin{aligned} \text{RP:}\quad & a_{11}x_{n+1} + a_{21}x_{n+2} + \ldots + a_{m1}x_{n+m} \geqslant h \\ & a_{12}x_{n+1} + a_{22}x_{n+2} + \ldots + a_{m2}x_{n+m} \geqslant h \\ & \quad\vdots \qquad\qquad\qquad \ldots \\ & a_{1n}x_{n+1} + a_{2n}x_{n+2} + \ldots + a_{mn}x_{n+m} \geqslant h \\ & x_{n+1} + x_{n+2} + \ldots + x_{n+m} = 1 \end{aligned} \tag{11.13}$$

and

$$x_{n+1}, x_{n+2}, \ldots x_{n+m} \geqslant 0$$

or in matrix form

$$\begin{aligned} & A'x \geqslant h.e_n \\ & x.e_m = 1, x \geqslant 0. \end{aligned} \tag{11.14}$$

One of course looks for the maximum value of the game h° and the corresponding optimal strategy. Assume that h^0 is positive; if this is not so one need only add suitably large positive quantity to the payoff matrix which makes the value of the game positive but does not change the optimal strategy given by the x vector. Define now a new set of variables

$$X_{n+i} = \frac{x_{n+i}}{h^0}, i = 1, 2, \ldots, m, \text{ then } X_{n+i} \geqslant 0 \text{ for } h^\circ > 0. \quad (11.15)$$

The row player's problem in (11.13) may now be rewritten as

$$\begin{aligned} a_{11}X_{n+1} + a_{21}X_{n+2} + \ldots + a_{ml}X_{n+m} &\geqslant 1 \\ a_{12}X_{n+1} + a_{22}X_{n+2} + \ldots + a_{m2}X_{n+m} &\geqslant 1 \\ \vdots \quad \vdots \qquad \vdots \qquad\qquad \vdots & \\ a_{1n}X_{n+1} + a_{2n}X_{n+2} + \ldots + a_{mn}X_{n+m} &\geqslant 1 \\ X_{n+1} + X_{n+2} + \ldots + X_{n+m} = \frac{1}{h^0} &= z \text{ say} \\ X_{n+1}, X_{n+2} + \ldots + X_{n+m} &\geqslant 0, \end{aligned} \quad (11.16)$$

where the maximum value of h^0 is sought. The maximum of h^0 implies minimum of z hence, by solving the above problem, viz.,

$$\min \underline{z} = X'e_n$$

$$\text{subject to } A^1X \geqslant e_n, X \geqslant 0, \quad (11.17)$$

the value of $h^0 = 1/\underline{z}$ is obtained which is also the value of the game. The optimal mixed strategy $x^\circ = h^0 . (X^0)$ is of course obtained by the relationship shown in (11.15).
In the same way the column player CP's problem may be transformed to the following linear program,

$$\begin{aligned} &\text{maximize} \quad \bar{z} = Y'e_n \\ &\text{subject to} \quad AY \leqslant e_m, Y \geqslant 0, \end{aligned} \quad (11.18)$$

where it is assumed that $g^0 > 0$, (g^0 is the value of the game) and the new variables Y_j are defined as,

$$Y_j = y_j/g^0, j = 1, 2, \ldots, n, Y_j \geqslant 0 \text{ for } g^0 > 0. \quad (11.19)$$

It is interesting to note that the Row and Column Player's problems put into the linear programming format (11.17), (11.18) represent a pair of problems dual to each other.

To transform a linear programming problem and its dual into a games theory format it is convenient to consider a problem in the inequality form. Given the problem,

$$Dx \leqslant d, x \geqslant 0, \qquad \text{maximize } \bar{z} = c'x, \quad (11.20)$$

and its dual

$$D'u \geqslant c, u \geqslant 0, \qquad \text{minimize } \underline{z} = d'u, \quad (11.21)$$

where D is the $m \times n$ constraint matrix, it follows from the duality properties considered in Chapter 6, that

$$c'x \leqslant d'u \quad \text{and} \quad \max \bar{z} = \min \underline{z} \tag{11.22}$$

Hence, if a feasible solution pair x^0, u^0 is found which further satisfies the inequality

$$c'x^0 - d'u^0 \geqslant 0, \tag{11.23}$$

then this in conjunction with (11.22) simply implies that

$$c'x^0 = d'u^0. \tag{11.24}$$

Thus x^0, u^0 must constitute an optimal solution pair. An $(m + n + 1)$ variable games problem may now be constructed by introducing a new scalar variable $t, 0 \leqslant t \leqslant 1$, and two vectors x, u such that,

$$X_j = t \,.\, x_j, \qquad U_i = t \,.\, u_i \tag{11.25}$$

and

$$X \,.\, e_n + U \,.\, e_m + t = 1\,. \tag{11.26}$$

From (11.25), (11.26) t may be expressed as

$$t = \frac{1}{x \,.\, e_n + u \,.\, e_m + 1}. \tag{11.27}$$

From (11.20), (11.21), (11.23), (11.25), the inequality system may be expressed as,

$$\begin{aligned} DX - td &\leqslant 0 \\ D'U - tc &\geqslant 0 \\ cX - dU &\geqslant 0. \end{aligned} \tag{11.28}$$

Changing the sign of the first inequality in (11.28), and adding (11.26) to the system we have

$$\begin{aligned} -DX + td &\geqslant 0 \\ D'U - tc &\geqslant 0 \\ cX - dU &\geqslant 0 \\ (X, U, t) \geqslant 0, \qquad Xe_n + Ue_m &+ t = 1. \end{aligned} \tag{11.29}$$

Compare the inequality system of (11.29) with the game having a skew symmetric pay-off matrix A,

$$A = \begin{bmatrix} 0 & -D & d \\ D' & 0 & -c \\ -d' & c' & 0 \end{bmatrix} \begin{matrix} m \\ n \\ 1, \end{matrix} \qquad (11.30)$$
$$\quad m \quad n \quad 1$$

and expressed as,

$$\begin{bmatrix} 0 & -D & d \\ D' & 0 & -c \\ -d' & c' & 0 \end{bmatrix} \times \begin{bmatrix} Y_1 \\ \\ Y_{m+n+1} \end{bmatrix} \geqslant M \,.\, e_{m+n+1} \qquad (11.31)$$

$$Y \geqslant 0,\ Y.\, e_{m+n+1} = 1, \text{ where, } Y = \{U, X, t\}.k$$

It is obvious from the *skew symmetry* of the game matrix that one person's loss is the other's gain and vice versa. The only numerical value that can make this possible is zero, hence this must be the value of the game; it also implies that the game is equivalent to the linear program stated above.

For the game specified by (11.31) the following property may be stated and proved.

If the symmetric game A has a solution for which $t > 0$, then both linear programming problems have solutions and conversely. Let

$$Y^0 = [U_1^0, U_2^0, \ldots, U_m^0, X_1^0, X_2^0, \ldots, X_n^0\ t], \qquad (11.32)$$

be an optimal strategy for the matrix game A with $t > 0$. Since the value w of the game is zero, we have $AY^\circ \geqslant 0$

or
$$\begin{aligned} -DX^0 + td &\geqslant 0 \\ D'U^0 - tc &\geqslant 0 \\ cX^0 - dU^0 &\geqslant 0, \end{aligned} \qquad (11.33)$$

and from the complementary property stated in (11.7) the last inequality must hold as equality (for $Y_{n+m+1} = t > 0$); hence $cX^0 = dU^0$.
From (11.33) one may write

$$\frac{DX^0}{t} \leqslant d \text{ or } Dx^0 \leqslant d \qquad (11.34)$$

$$\frac{D'U^0}{t} \geqslant c \text{ or } D'u^0 \geqslant c$$

and
$$cx^0 = du^0, \qquad (11.35)$$

where $x^0 = (X^0/t)$, $u^0 = (U^0/t)$ and (x^0, u^0) constitute optimal feasible solutions to the primal and dual problems respectively; optimality is of course assured by the relationship (11.35).

The converse relationship may be similarly deduced and is left as an exercise for the reader.

11.4 MATRIX GAMES AND LINEAR PROGRAMS AN EXAMPLE

Consider the matrix game specified by the pay-off matrix A,

$$A = \begin{bmatrix} 3 & 4 & 8 \\ 6 & 5 & 4 \end{bmatrix} \tag{11.36}$$

which has not got an equilibrium point in terms of a pure strategy (this has been already discussed in section 11.2). To investigate the strategies of each player the problem of each player may be expressed as

$$\begin{aligned} \text{RP:} \quad & 3x_4 + 6x_5 \geqslant h \\ & 4x_4 + 5x_5 \geqslant h \\ & 8x_4 + 4x_5 \geqslant h \\ & x_4 + x_5 = 1, x_4, x_5 \geqslant 0 \end{aligned} \tag{11.37}$$

and,

$$\begin{aligned} \text{CP:} \quad & 3y_1 + 4y_2 + 8y_3 \leqslant g \\ & 6y_1 + 5y_2 + 4y_3 \leqslant g \\ & y_1 + y_2 + y_3 = 1, \qquad y_1, y_2, y_3 \geqslant 0. \end{aligned} \tag{11.38}$$

Introducing surplus variables $x_1, x_2, x_3 \geqslant 0$, and slack variables $y_4, y_5 \geqslant 0$, these may be rewritten as,

$$\begin{aligned} x_1 &= 3x_4 + 6x_5 - h \\ x_2 &= 3x_4 + 5x_5 - h \\ x_3 &= 8x_4 + 4x_5 - h \\ 1 &= x_4 + x_5 + 0, \\ & x_1, x_2, \ldots, x_5 \geqslant 0, \end{aligned} \tag{11.39}$$

$$\begin{aligned} -y_4 &= 3y_1 + 4y_2 + 8y_3 - g \\ -y_5 &= 6y_1 + 5y_2 + 4y_3 - g \\ 1 &= y_1 + y_2 + y_3 + 0, \\ & y_1, y_2, \ldots, y_5 \geqslant 0. \end{aligned} \tag{11.40}$$

As in the contracted tableau representation for linear programs, these can be expressed by the following schema (note the negative variables occur in the right hand side of the schema rather than across the schema).

	y_1	y_2	y_3	g	
$-y_4 =$	3	4	8	-1	x_4
$-y_5 =$	6	5	4	-1	x_5
$-1 =$	-1	-1	-1	0	h
	$=x_1$	$=x_2$	$=x_3$	$=-1$	

SCHEMA 11.0

Of all the basic solutions possible to this system, $5!/3!2! = 10$, this particular one is inadmissible since $x_4, x_5 = 0$ (in fact none of the variables have non zero value) and hence this does not satisfy the last equation of (11.39). Therefore, first rewrite x_5 in terms of x_1 and x_4 and substitute in rest of the equations in (11.39):

$$\begin{aligned} x_5 &= 1/6h - 3/6x_4 + \tfrac{1}{6}x_1, \\ x_2 &= -1/6h + 9/6x_4 + \tfrac{5}{6}x_1, \\ x_3 &= -2/6h + 6x_4 + \tfrac{4}{6}x_1, \\ 1 &= 1/6h + 3/6x_4 + \tfrac{1}{6}x_1 \end{aligned} \tag{11.41}$$

Rewrite the last equation as $h = 6 - 3x_4 - x_1$ and substitute it in the rest to obtain,

$$\begin{aligned} x_5 &= 1 - x_4 + 0, \\ x_2 &= -1 + 2x_4 + x_1, \\ x_3 &= -2 + 7x_4 + x_1, \\ h &= 6 - 3x_4 - x_1. \end{aligned} \tag{11.42}$$

This can be expressed in the form of Schema (11.1) to be read columnwise. Because of the skew symmetry property of the system the same schema considered row-wise represents the column player's problem (CP) similarly transformed.

	1	y_5	y_2	y_3	
$-y_4 =$	-3	-1	2	7	x_4
$-y_1 =$	-1	0	1	1	x_1
$g =$	6	1	-1	-2	1
	$=h$	$=x_5$	$=x_2$	$=x_3$	

SCHEMA 11.1

Schemas 11.1–11.9 illustrate all the basic solutions and hence strategies in the game. Let $\{p, q\}$, $\{r, s, t\}$ represent the indices of the y variables and x variables respectively which are basic in these solutions. If one observes the Table 11.2 which summarizes the pertinent information concerning the game, it is found $p = 2$, $q = 3$, $y_2 = 4/5$, $y_3 = 1/5$, $r = 1$, $s = 4$, $t = 5$, $x_1 = 3/5$, $x_4 = \frac{1}{5}$, $x_5 = 4/5$ (Schema 11.9) contains the optimal solution to the game of value $h = g = w = 24/5$. To relate the obvious connection between solving a game problem and the simplex method of exploring the adjacent vertices of a convex polyhedron, observe Fig. 11.1: A network of 9 connected nodes

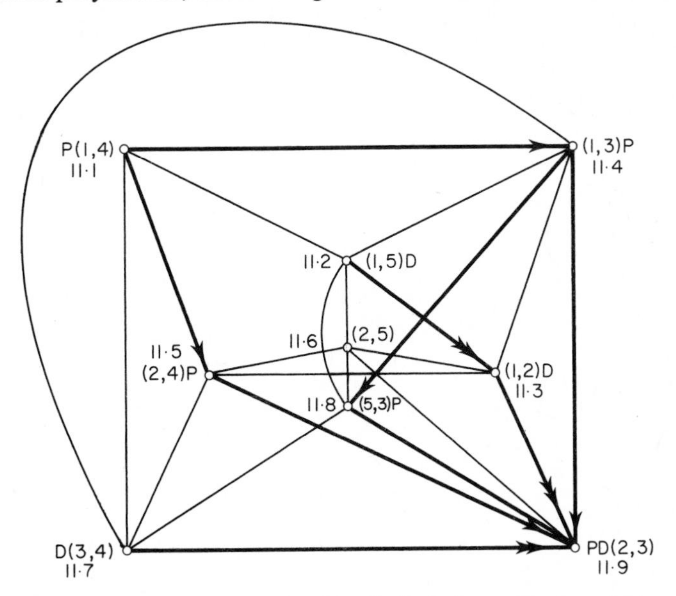

P=primal feasible. D=dual feasible. (I, J)=indices of basic variables.

FIG. 11.1

represents the nine Schema 11.1–11.9. If we assume that the problem stated in y is the primal problem and that in x the dual problem, then the path of the primal simplex method to optimum is shown with single arrows and those with double arrows indicate the dual simplex paths. Note that some paths may not be either, these are shown with single line.

Finally, the column player's problem may be transformed into a linear program and stated as,

$$3Y_1 + 4Y_2 + 8Y_3 \leqslant 1$$

$$6Y_1 + 5Y_2 + 4Y_3 \leqslant 1 \tag{11.43}$$

Maximize $Y_1 + Y_2 + Y_3 = 1/g$ where $Y_j = (y_j/g) \geqslant 0, j = 1, 2, 3$ and $g > 0$

(1, 4)	1	y_5	y_2	y_3	
$-y_4 =$	−3	−1	2	7	x_4
$-y_1 =$	−1	0	1	1	x_1
$g =$	6	1	−1	−2	1
	$= h$	$= x_5$	$= x_2$	$= x_3$	

SCHEMA 11.1

(1, 5)	1	y_4	y_2	y_3
$-y_5 =$	3	−1	−2	−7
$-y_1 =$	−1	0	1	1
$g =$	3	1	1	5
	$= h$	$= x_4$	$= x_2$	$= x_3$

SCHEMA 11.2

(1, 2)	1	y_5	y_4	y_3
$-y_2 =$	−3/2	−1/2	1/2	7/2
$-y_1 =$	1/2	1/2	−1/2	−5/2
$g =$	9/2	1/2	1/2	3/2
	$= h$	$= x_5$	$= x_4$	$= x_3$

SCHEME 11.3

(1, 3)	1	y_5	y_2	y_4
$-y_3 =$	−3/7	−1/7	2/7	1/7
$-y_1 =$	−4/7	1/7	5/7	−1/7
$g =$	36/7	5/7	−3/7	2/7
	$= h$	$= x_4$	$= x_2$	$= x_4$

SCHEMA 11.4

(2, 4)	1	y_5	y_1	y_2
$-y_4 =$	−1	−1	−2	5
$-y_3 =$	−1	0	1	1
$g =$	5	1	1	−1
	$= h$	$= x_4$	$= x_1$	$= x_2$

SCHEMA 11.5

(2, 5)	1	y_4	y_1	y_3
$-y_5 =$	1	−1	2	−5
$-y_2 =$	−1	0	1	1
$g =$	4	1	−1	4
	$= h$	$= x_4$	$= x_1$	$= x_3$

SCHEMA 11.6

(3, 4)	1	y_5	y_2	y_1
$-y_4 =$	4	−1	−5	−7
$-y_3 =$	−1	0	1	1
$g =$	4	1	1	2
	$= h$	$= x_5$	$= x_2$	$= x_1$

SCHEMA 11.7

(3, 5)	1	y_4	y_2	y_1
$-y_5 =$	−4	−1	5	7
$-y_3 =$	−1	0	1	1
$g =$	8	1	−4	−5
	$= h$	$= x_4$	$= x_2$	$= x_1$

SCHEMA 11.8

(2, 3)	1	y_5	y_1	y_4	
$-y_3 =$	−1/5	−1/5	2/5	1/5	x_2
$-y_2 =$	−4/5	1/5	7/5	−1/5	x_3
$g =$	24/5	4/5	3/5	1/5	1
	$= h$	$= x_5$	$= x_1$	$= x_4$	

SCHEMA 11.9

		$-Y_1$	$-Y_2$	$-Y_3$
$1/g_0$	0	−1	−1	−1
Y_4	1	3	4	8
Y_5	1	6	5	4

TABLEAU 11.10

		$-Y_5$	$-Y_2$	$-Y_3$
$1/g_0$	1/6	1/6	−1/6	−2/6
Y_4	3/6	−3/6	9/6	36/6
Y_1	1/6	1/6	5/6	4/6

TABLEAU 11.11

		$-Y_5$	$-Y_2$	$-Y_4$
$1/g_0$	7/36	5/36	−1/12	1/18
Y_3	1/12	−1/12	1/4	1/6
Y_1	1/9	2/9	4/6	−1/9

TABLEAU 11.12

		$-Y_5$	$-Y_1$	$-Y_4$
$1/g$	5/24	1/6	1/8	1/24
Y_3	1/24	−1/6	−3/8	5/24
Y_2	1/6	1/3	6/4	−1/6

TABLEAU 11.13

Complementary Basic Solutions for the Games Problem

(p, q)	h	x_1	x_2	x_3	x_4	x_5	Feasible?	(r, s, t)	g	y_1	y_2	y_3	y_4	y_5	Feasible?
(14)	6	0	−1	−2	0	1	No	(235)	6	1	0	0	3	0	Yes
(15)	3	0	1	5	1	0	Yes	(234)	3	1	0	0	0	−3	No
(12)	9/2	0	0	3/2	$\frac{1}{2}$	−1/2	No	(345)	9/2	−1/2	3/2	0	0	0	No
(13)	36/7	0	−3/7	0	2/7	5/7	No	(245)	36/7	4/7	0	3/7	0	0	Yes
(24)	5	1	0	−1	0	1	No	(135)	5	0	1	0	1	0	Yes
(25)	4	−1	0	4	1	0	No	(134)	4	0	1	0	0	−1	No
(34)	4	2	1	0	0	1	Yes	(125)	4	0	0	1	−4	0	No
(35)	8	−5	−4	0	1	0	No	(124)	8	0	0	1	0	4	Yes
(23)	24/5	3/5	0	0	1/5	4/5	Yes*	(145)	24/5	0	4/5	1/5	0	0	Yes*

* Both the inequality systems have feasible solutions hence this is the optimal solution.

TABLE 11.2

Introduce slacks Y_4, Y_5 and express in the form of a contracted tableau, Tableau 11.10. This is immediately comparable with Schema 11.0. The exchange Y_1/Y_5 leads to Tableau 1.11 which compares (in solution values) with Schema 11.1 when multiplied by $g = 6(1/g = 1/6)$. After two further exchanges the optimal tableau, Tableau 11.13 is is obtained where $1/g = 5/24$, hence, the value of the game is 24/5 and the optimal strategy is

$$y_3 = g \,.\, Y_3 = \frac{24}{5} \cdot \frac{1}{24} = 1/5 \text{ and } y_2 = g \,.\, Y_2 = 3/5.$$

Again Tableau 11.13 compares directly with the optimal schema 11.9 of the Games Problem.

11.5 STATEMENT OF LINEAR PROGRAM, QUADRATIC PROGRAM, AND THE MATRIX GAME AS A FUNDAMENTAL PROBLEM [33] [13]

In this section various programming and game problems are expressed in the format of another problem which is called the "Fundamental Problem".

Fundamental problem

Given the square matrix M of dimension $(m + n) \times (m + n)$ and the vector q of dimension $(m + n)$ find the two non-negative vectors w, z, each of dimension $(m + n)$ such that they solve the system,

$$w = q + Mz, \tag{11.44}$$

satisfy the non-negative relations,

$$w \geqslant 0, z \geqslant 0 \tag{11.45}$$

and the complementary relation,

$$w'z = 0. \tag{11.46}$$

In stating these problems in the form of the Fundamental Problem some properties of the matrix M will follow; these are referred to in the relevant parts of this section.

Linear program

Consider the linear programming problem,

$$\begin{array}{ll} \text{Maximize} & c'x \\ \text{subject to} & Ax \leqslant b, x \geqslant 0, \end{array} \tag{11.47}$$

and its dual

$$\begin{array}{ll} \text{Minimize} & b'v \\ \text{subject to} & A'v \geqslant c, v \geqslant 0, \end{array} \tag{11.48}$$

where A is an $m \times n$ matrix, c, x are n-vectors and b, v are m-vectors. Introducing m-dimensional vector y of slack variables and n-dimensional vector u of surplus variables these problems may be re-expressed as:

$$\begin{array}{lll} \text{Maximize} & c'x & \\ \text{subject to} & Ax + Iy = b, & x, y \geqslant 0, \end{array}$$

and

$$\begin{array}{lll} \text{minimize} & b'v & \\ \text{subject to} & A'v - Iu = c, & u, v \geqslant 0 \end{array} \tag{11.49}$$

From the duality theory of linear programming it follows that for the optimum feasible solutions to this problem pair the following relationship must hold,

$$\begin{pmatrix} u \\ y \end{pmatrix} = \begin{pmatrix} -c \\ b \end{pmatrix} + \begin{pmatrix} 0 & A' \\ -A & 0 \end{pmatrix} \begin{pmatrix} x \\ v \end{pmatrix}, \qquad x, y, v, u \geqslant 0 \tag{11.50}$$

and $x'u + y'v = 0$.

If one substitutes

$$w = \begin{pmatrix} u \\ y \end{pmatrix}, \quad q = \begin{pmatrix} -c \\ b \end{pmatrix}, \quad z = \begin{pmatrix} x \\ v \end{pmatrix}, \quad M = \begin{pmatrix} 0 & A' \\ -A & 0 \end{pmatrix}$$

then this becomes equivalent to the fundamental problem.

Quadratic programming

Consider the symmetric quadratic programming problem in the general form (due to Cottle [13]) viz.,
Find vectors x, v and minimum $f(x, v)$ such that

$$Ax + Ev \geqslant b, x \geqslant 0, f(x, v) = c'x + \tfrac{1}{2}(x'Dx + v'Ev) \tag{11.51}$$

This has associated with it the dual problem:

Find $\hat{v}$ and maximum $G(\hat{x}, \hat{v})$ such that

$$-D\hat{x} + A'\hat{v} \leqslant c \quad \text{and} \quad \hat{v} \geqslant 0, \qquad G(\hat{x}, \hat{v}) = b\hat{v} - \tfrac{1}{2}(\hat{x}D\hat{x} + \hat{v}E\hat{v}) \tag{11.52}$$

For $E = 0$ the simple primal quadratic programming problem as stated in

Chapter 10 is obtained, the corresponding dual has been stated by Dorn, [19]. Further, for $D = 0$ the problem and its dual reduce to a pair of primal dual linear programs.

If the necessary relationships of Kuhn–Tucker for the optimal solution are set out then a system directly comparable with the Fundamental Problem follows. This deduction is now left as an exercise for the reader.

Hint:

$$M = \begin{bmatrix} D & -A' \\ A & E \end{bmatrix}, \quad q = \begin{pmatrix} c \\ -b \end{pmatrix} \text{etc.} \tag{11.53}$$

Note that if D and E are positive semi-definite matrices then so is M and this ensures that any local optimum solution to the primal or dual problem is also globally optimum.

Bimatrix game

The problem of simple matrix game may be transformed into the fundamental problem in the same way as one transforms the associated linear program into this form.

Consider now the bimatrix game (see Section 11.2) defined by the two pay-off matrices A, B each of dimension $m \times n$. It follows from the necessary conditions for an equilibrium point that,

$$\begin{aligned} y &= Ax - e_m \quad y,\ x \geqslant 0 \\ u &= B'v - e_n \quad u,\ v \geqslant 0 \end{aligned} \tag{11.54}$$

and $xu + yv = 0$.

This is once again in the form of the Fundamental Problem, where

$$M = \begin{pmatrix} 0 & A \\ B' & 0 \end{pmatrix}, \quad w = \begin{pmatrix} y \\ u \end{pmatrix}, \quad q = \begin{pmatrix} -e_m \\ -e_n \end{pmatrix}, \quad z = \begin{pmatrix} v \\ x \end{pmatrix}. \tag{11.55}$$

For $A > 0$, $B > 0$, the matrix M is “copositive”. A square matrix M is copositive if $z \geqslant 0$ implies $zMz \geqslant 0$. Of the most important class of such matrices are the positive semi-definite matrices, $zMz \geqslant 0$ for all z and non-negative matrices viz $M \geqslant 0$ which indicates each element of the matrix is non-negative.

11.6 COMPLEMENTARY PIVOT THEORY: LEMKE'S ITERATIVE METHOD, AND DANTZIG AND COTTLE'S PRINCIPLE PIVOT METHOD FOR SOLVING THE FUNDAMENTAL PROBLEM

Philip Wolfe [55] in his rather ad hoc method of solving Quadratic Programs was the first person to use pivot rules which satisfied complementary conditions. Since then Lemke [33] and Dantzig and Cottle [13] have devised pivoting algorithms which are designed to satisfy feasibility and complementarity properties and hence solve the fundamental problem. The applicability and the convergence of such algorithms depend on the properties of the M matrix in (11.44).

Lemke's iterative method

In this method by introducing a pseudo variable (artifical variable with no formal penalty) with a column of all positive coefficients a basic feasible solution is constructed as a starting point for an augmented problem. A subsequent sequence of feasible pivots then leads to satisfy the complementarity conditions. For a formal description of the method it is necessary to introduce the following terminology.

A solution to (11.44) is feasible if and only if all the variables are non-negative, viz., $w \geqslant 0$, $z \geqslant 0$. Assume for simplicity of exposition that all the basic solutions are non-degenerate. Hence, any basic solution with exactly $m + n$ of the w and z positive and the other $m + n$ variables set to zero is a basic feasible solution. Let L denote the set of all solutions of (11.44) and K the set of all feasible solutions and C the set of all the feasible solutions which also satisfy complementarity condition; K and C may be empty.

Given a system

$$w = q + Az, \tag{11.56}$$

where A has $(m + n)$ rows, but may have more than $m + n$ columns, the transformed system may be expressed as

$$w^t = q^t + A^t z^t, \tag{11.57}$$

where the transformation mentioned is a simple exchange, i.e., a pivotal transformation; (w^t, z^t) is some permutation of (w, z); q^t, A^t are the transformed q and A, t the step of the transformation. For $q^t > 0$, the system in (11.57) represents a "basic feasible" (point) solution: this is also an extreme point of K. A "feasible pivot" is a pivot from one feasible basic form to another. A variable z_j^t currently nonbasic may be made basic and a finite basic feasible solution obtained if at least one $a_{sj}^t < 0$.

If the w_s^t/z_j^t exchange defines one such pivot step then this leads to a finite increase in the value of z_j^t along a bounded edge of $K: 0 \leqslant z_j^t \leqslant |q_s^t/a_{sj}^t|$. Two basic feasible sets separated from each other by one feasible pivot are "adjacent" and differ from each other in exactly one variable making up the basic sets.

For $a_j^t \geqslant 0$, however, an unbounded edge of K is detected and increasing value of z_j^t generates a ray. A "feasible pivotal scheme" consists of a sequence of feasible pivots, and defines an adjacent extreme point path. Finally, a "proper pivotal" scheme is defined as one for which no basic set appears twice; the number of basic sets being finite such a scheme always terminates.

Starting with the fundamental problem (11.44) augment the problem by a pseudo variable z_0 and a column e of unit components to obtain the system

$$w = q + z_0 e + Mz = q + A\bar{z}, \tag{11.58}$$

where the vector $\bar{z}$ making up the nonbasic set has $m + n + 1$ components

$$\bar{z} = \left(\frac{z_0}{z}\right) \quad \text{and} \quad A = [e, M].$$

Let L' denote the set of all solutions to (11.58) and let K' denote the set of feasible solutions, i.e., w, z, $\geqslant 0$ and let C_0 denote the extreme points of K' which satisfy the condition

$$w_i z_i = 0, \qquad i = 1, 2, \ldots, n + m. \tag{11.59}$$

If all q_i are positive then the problem is trivially solved, the initial solution being both feasible and complementary. Otherwise for some $q_r < 0$ carry out the exchange w_r/z_0 and obtain the system,

$$w^t = q^t + A^t \bar{z}^t, \qquad q^t > 0. \tag{11.60}$$

In this observe that the relationship (11.59) is satisfied. One pair of variables (w_r, z_r) is non basic and a basic feasible solution of (11.60) is not a basic feasible solution of (11.44) as long as z_0 remains basic.

For $t = 1$, w_r has just become non-basic and the corresponding column a_1^t has all unit entries. This simply implies that there exists a corresponding unbounded edge (increase w_r to $\rightarrow \alpha$); label this edge E_0. Now in any subsequent step always try to bring into the basis the appropriate member of the (w_s, z_s) non basic pair and maintain feasibility. If the terminal basic forms ($t = 1$ or t corresponding to the step in which full complementarity is achieved) are left out then in all the intermediate steps there are exactly two adjacent basic forms satisfying (11.59). The sequence may terminate in one of two ways.

(i) The basic variable z_0 becomes non basic and this immediately makes the corresponding basic solution also a basic solution in C, (11.44), (11.45), (11.46).

(ii) The pivot sequence ends in an unbounded edge E of K'; this implies that for M a copositive plus matrix no solution satisfying the desired conditions can exist (for proof see [33]).
[A copositive plus matrix is a copositive matrix for which $z \geqslant 0$ and $z \neq 0$ implies $zMz > 0$; the terminology is due to Cottle.]

The principal pivoting method of Dantzig and Cottle

This method predates the method of Lemke and in many ways has close similarity with the latter. No artificial variables are needed in this to make the starting solution feasible; instead variables are admitted at negative level, complementarity is maintained for all but one pair of variables. In a sequence of pivots through one almost complementary path (see Chapter 10), one such negative variable is removed.

There are two cycles in the algorithm; the major cycle each of which removes one negative variable and the minor cycle in which a sequence of almost complementary basis is generated.

Step. O. Major cycle

Start with the basic solution $(w, z) = (q^t, 0)$. If $q^t \geqslant 0$ for all the components then the procedure terminates. Otherwise identify a variable $w_s = q_s^t < 0$ and carry out the following steps. (Note in these steps $z_i w_i = 0$ for $i \neq s$).

Step. 1. *Minor cycle*

Increase z_s^t until it is blocked by a positive basic variable dropping to zero or w_s^t increasing to zero.

Step 2.

Make the blocking variable non basic by pivoting its complement into the basic set. A major cycle is completed if w_s^t drops out of the basic set, otherwise go to Step I and carry out another Minor Cycle.

The principal pivoting method terminates in a solution of the fundamental problem if M has positive principal minors (and, in particular, if M is positive definite). With slight modification the same algorithm may be restated for the case where M is positive semi-definite, hence both quadratic and linear programs may be solved by this method.

EXERCISES

11.1 Given the payoff matrix

$$\begin{bmatrix} 7 & 8 & 12 \\ 10 & 9 & 8 \end{bmatrix},$$

corresponding to a two person zero sum game, find the optimal mixed stragey for both players. What is the equivalent linear programming formulation of the problem?

11.2 Apply Lemke's method to solve the problem

$$w_1 = 10 - 2z_1 + 3z_2 - z_3$$
$$w_2 = -1 + z_1 + 2z_2 + z_3$$
$$w_3 = 3 - z_1 + 2z_2 + 3z_3,$$

such that w_1, w_2, w_3 and $z_1, z_2, z_3 \geqslant 0$, and $w_1z_1 + w_2z_2 + w_3z_3 = 0$.

11.3 Formulate the problems in exercise 10.2 and 10.3 as linear complementarity problems. Comment on the applicability of the algorithms discussed in Chapter 11 to solve these problems.

References

1. Arrow, K. J., Uzawa, H. and Hurwicz, L., "Studies in Linear and Nonlinear Programming", Stanford University Press, 1958.
2. Beale, E. M. L., "Mathematical Programming in Practice", Pitman, 1968.
3. Beale, E. M. L., "Advanced Algorithmic Facilities for Mathematical Programming Systems, Integer and Nonlinear Programming", Edited by J. Abadie, North Holland, 1969.
4. Beale, E. M. L., 'Sparseness in linear programming', presented to Oxford Symposium on Sparse Systems of Linear Equations, 1970.
5. Beale, E. M. L. and Small, R. E., Mixed integer programming by a branch and bound technique. *In* "Proceedings of the IFIP Congress, 1965", Edited by W. A. Kalerick, Spartan Press, New York.
6. Beale, E. M. L. and Tomlin, J. A., Special facilities in a general mathematical programming system for non-convex problems using special sets of variables. *In* "Proceedings 5th IFORS Conference", Wiley, New York, 1969.
7. Bellman, R. E. and Dreyfus, S. E., "Applied Dynamic Programming", Princeton University Press, 1963.
8. Benichou, M., *et al.,* Experiments in mixed-integer linear programming, *Mathematical Programming,* **1** (1971), 76–94.
9. Bracken, J. and McCormick, G. P., "Selected Applications of Nonlinear Programming", John Wiley & Sons, 1968, p. 94–100.
10. Brayton, R. K., Gustavson, F. G. and Willoughby, R. A., "Some Results on Sparse Matrices", RC-2332, IBM Research Centre, York Town Heights, 1969.
11. Charnes, A., Cooper, W. W. and Henderson, A., "An Introduction to Linear Programming", John Wiley & Sons, 1953.
12. Charnes, A. and Cooper, W. W., Chance constraints and normal deviates, *Journal of the American Statistical Association,* **57**, No 297, 135–149.
13. Cottle, R. W. and Dantzig, G. B., Complementary pivot theory of mathematical programming. *In* "Mathematics of Decision Sciences", published by the American Mathematical Society, 1968, p. 115–136.
14. Dakin, R. J., A tree search algorithm for mixed integer programming problems, *Computer Journal,* **8** (1965), 250–255.
15. Dantzig, G. B., "Linear Programming and Extensions", Princeton University Press, 1963.
16. Dantzig, G. B. and Van-Slyke, R. M., "Generalized Upper Bounding Techniques for Linear Programming", ORC64-17, 64–18, University of California, OR Centre report, 1964.

17. Dantzig, G. B. and Veinott, A. F. (Eds), "Mathematics of Decision Sciences", Published by the American Mathematical Society, 1968.
18. Driebeek, N. J., An algorithm for the solution of mixed integer programming problems, *Man. Sci.*, **12** (1966), 7–15.
19. Dorn, W. S., Duality in quadratic programming, *Q. Appl. Math.*, **18** (1960), 155–162.
20. Duval, P., The unloading problem of plane curves, *Amer. J. Math.* **62** (1940), 307–311.
21. Fiacco, A. V. and McCormick, G. P., "Nonlinear Programming, Sequential Unconstrained Minimization Techniques", Research Analytic Corporation (RAC) Research Series, John Wiley & Sons, 1968.
22. Ford, L. R. and Fulkerson, R., "Flows in Networks", Princeton University Press, 1962.
23. Forsythe, G. and Moler, C. B., "Computer Solution of Linear Algebraic Equations", Prentice Hall, 1967.
24. Gomory, R. E., An algorithm for integer solution to linear programs. *In* "Recent Advances in Mathematical Programming", Edited by R. Graves and P. Wolfe, McGraw-Hill, 1963, p. 269–302.
25. Gomory, R. E., "An All Integer Integer Programming Algorithm, Industrial Scheduling", Edited by J. Muth and G. L. Thompson, Prentice Hall, 1963.
26. Hadley, G. F., "Linear Programming", Addison Wesley, Reading, Massachusetts, 1962.
27. Hadley, G. F., "Nonlinear and Dynamic Programming", Addison Wesley, Reading, Massachusetts, 1964.
28. Hancock, H., "Theory of Maxima and Minima", Ginn & Co., 1917, Dover Publications, 1960.
29. Healy, W. C., "Multiple Choice Programming", *Operations Research,* **12** (1964), 122–138.
30. Kuhn, H. W. and Tucker, A. W., Nonlinear programming. *In* "Proceedings of the Second Berkley Symposium on Mathematical Statistics and Probability", 1951, p. 481–492.
31. Kuhn, H. W., (Ed.), "Proceedings of the Princeton Symposium on Mathematical Programming", Princeton University Press, 1970, p. 179–323.
32. Land, A. H. and Doig, A. G., An automatic method of solving discrete programming problems, *Econometrica,* **28**, (1960), 497–520.
33. Lemke, C. F., On Complementary Pivot Theory, in Ref. 17, p. 95–114.
34. Lowes, S. L., Boot, J. C. G. and Wage, S., A quadratic programming approach to the problem of the optimal use of milk in the Netherlands, *Journal of Farm Economics*, **45** (1963), 309–317.
35. Luce, R. D. and Raiffa, H., "Games and Decisions, Introduction and Critical Survey", John Wiley & Sons, New York, 1957.
36. Miller, C. E., The simplex method of local separable programming. *In* "Recent Advances in Mathematical Programming" (see Ref. 24).
37. Mitra, G. and Wolfenden, K., A computer technique for optimizing the sites and heights of transmission line towers: A dynamic programming approach, *Computer Journal*, **10**, No. 4 (1968).
38. Mitra, G., Designing branch and bound algorithms for mathematical program-

ming. Presented to the 7th International Symposium on Mathematical Programming, The Hague, 1970, also *Mathematical Programming,* **4** (1973), 155–170.

39. Mitra, G., Richards, D. B. C. and Wolfenden, K., An improved algorithm for the solution of integer programs by the solutions of associated diophantine equations, *RIRO*, **4**, No. R-1 (1970), 47–60.
40. Mitra, G., Sparse inverse in the factored form and maintaining sparsity during simplex iterations. *In* "Software for Numerical Mathematics, Edited by D. J. Evans, Academic Press, 1974, p. 85–97.
41. Mitra, G., A Dichotomizing Procedure for the Fixed Charge Problem, Presented to SIAM conference, Toronto, 1968.
42. Mitra, G. and Calogero, V., A computer method for optimizing the vertical profile of highways: A mathematical programming approach. *In* "Proceedings of the Conference on Computer Aided Design", I.E.E., Southampton, 1959.
43. Orchard-Hays, W., "Advanced Linear Programming Computing Techniques," McGraw-Hill, 1969.
44. Proschan, F. and Bray, T. A., Optimal redundancy under multiple constraints, *Operations Research,* **13** (1964), 800–814.
45. Simonnard, M. A., "Linear Programming", translated by W. S. Jewell, Prentice Hall, 1966.
46. Stiefel, E. L., "An Introduction to Numerical Mathematics", Academic Press, 1963.
47. Tomlin, J. A., "Maintaining a Sparse Inverse in the Simplex Method", Technical Report No. 70-16, Nov. 1970, OR House, Stanford University, California.
48. Tomlin, J. A. and Forrest, J. J. H., Updating triangular factors of the basis to maintain sparsity in the product form of inverse simplex method, *Mathematical Programming Journal* (1972).
49. UMPIRE, Unified Mathematical Programming Incorporating Refinements and Extensions, User Manual, Scientific Control Systems UK, Ltd.
50. Van-de-Panne, C. and Popp, W., Minimum cost cattle feed under probabilistic protein constraints, *Management Science,* **9** (1963), 405–430.
51. Van-de-Panne, C. and Whinston, A., The simplex and dual method for quadratic programming, *OR Quarterly*, **15**, No. 4, Dec. 1964.
52. Vajda, S., "Mathematical Programming", Addison Wesley, Reading, Massachusetts, 1961.
53. Von-Neumann, J. and Morgenstern, O., "Theory of Games and Economic Behaviour", Princeton University Press, Ed. 1, (1944), Ed. 2, (1947).
54. Wagner, H. M., "Principles of Operations Research," Prentice Hall, 1969.
55. Wolfe, P., The simplex method for quadratic programming, *Econometrica,* **27** (1959), 382–398.
56. Wolfe, P., The composite simplex algorithm, *SIAM Review*, **7**, No. 1, (1965), 42–54.

APPENDIX 1

Some Mathematical Background

A1.1 LINEAR DEPENDENCE OF VECTORS AND RANK OF A MATRIX

A vector a_t from the n-dimensional euclidean space E^n is said to be a linear combination of the vectors $a_1, a_2, \ldots, a_k$ in E^n if the following equality

$$a_t = \lambda_1 a_1 + \lambda_2 a_2 + \lambda_2 a_2 \ldots + \lambda_k a_k, \tag{A1.1}$$

holds for a set of scalars $\lambda_1, \lambda_2, \ldots, \lambda_k$.

A set of vectors $a_1, a_2, \ldots, a_p$ from E^n is said to be linearly dependent if there exist scalars λ_i with at least one or more non zero such that

$$\lambda_1 a_1 + \lambda_2 a_2 + \ldots + \lambda_p a_p = 0. \tag{A.1.2}$$

If (A1.2) holds only for the case $\lambda_1 = \lambda_2 = \ldots = \lambda_p = 0$ then the vectors are said to be linearly independent. Thus by definition if a set of vectors is not linearly dependent then the vectors are linearly independent. However, a vector a_t is said to be linearly dependent on the set of vectors $a_1, a_2, \ldots, a_k$ when the equality (A.1.1) holds for a set of non-zero scalars $\lambda_1, \lambda_2, \ldots, \lambda_k$. In this context it must be emphasized that linear dependence and linear independence are properties applying to a set of vectors and not to an individual vector.

Rank of a matrix.

Let A be a matrix of dimension $m \times n$. Then the n columns of A may be thought of as a set of n vectors in E^m, since each column is an m component vector. The rank or the column rank of A may be denoted by rank (A) and is defined to be the largest number of linearly independent columns in A. It can be shown [A1] that the rank of A is k if and only if every minor of A of order $k + 1$ vanishes while there is at least one minor of order k that does not vanish. From this two conclusions follow.

(a) For $m \leqslant n$ as in linear programming constraint matrices the rank of A can only be m or less.
(b) The column and row rank of a matrix are identical.

A.1.2 BASIC SOLUTION TO AN EQUALITY SYSTEM AND GRAPHICAL REPRESENTATION

The concept of a basis is very important in linear programming; prior to defining this it is convenient to define a spanning set.

Spanning set

A set of vectors $a_1, a_2, \ldots, a_k$ taken from E^m is said to span E^m if every vector in E^m can be written as a linear combination of $a_1, a_2, \ldots, a_k$.

The unit vectors,

$$e_1 = \begin{bmatrix} 1 \\ 0 \\ 0 \\ \cdot \\ \cdot \\ 0 \\ 0 \end{bmatrix}, \quad e_2 = \begin{bmatrix} 0 \\ 1 \\ \cdot \\ \cdot \\ \cdot \\ 0 \\ 0 \end{bmatrix}, \ldots, e_m = \begin{bmatrix} 0 \\ 0 \\ \cdot \\ \cdot \\ \cdot \\ 0 \\ 1 \end{bmatrix},$$

span E^m, for the scalar components of any vector a_t in E^m may multiply each of these to obtain a representation of the vector a_t itself.

Basis

A basis for E^m is a linearly independent subset of vectors from E^m which spans all of E^m.

A few related properties may now be stated.

(i) The representation (by multipliers) of any vector in E^m by a set of basis vectors is unique.
(ii) The number of vectors forming a basis in E^m is exactly m.

Some of the concepts associated with a basis are now illustrated by examples in E^3. Consider the set of inequalities,

$$\begin{aligned} a_{11}x_1 + a_{12}x_2 + a_{13}x_3 &\leqslant b_1 \\ a_{21}x_1 + a_{22}x_2 + a_{23}x_3 &\leqslant b_2 \\ a_{31}x_1 + a_{32}x_2 + a_{33}x_3 &\leqslant b_3 \end{aligned}$$

and

$$x_1, x_2, x_3 \geqslant 0.$$

Let numerical values be assigned to these vectors,

$$a_1 = \begin{bmatrix} a_{11} \\ a_{21} \\ a_{31} \end{bmatrix} = \begin{bmatrix} 2 \\ 1 \\ 3 \end{bmatrix}, \quad a_2 = \begin{bmatrix} a_{12} \\ a_{22} \\ a_{32} \end{bmatrix} = \begin{bmatrix} 1 \\ 2 \\ 1 \end{bmatrix}, \quad a_3 = \begin{bmatrix} a_{13} \\ a_{23} \\ a_{33} \end{bmatrix} = \begin{bmatrix} 1 \\ 1 \\ 1 \end{bmatrix},$$

$$b = \begin{bmatrix} b_1 \\ b_2 \\ b_3 \end{bmatrix} = \begin{bmatrix} 6 \\ 7 \\ 8 \end{bmatrix} \tag{A.1.3}$$

Now consider the two spaces,

Solution Space E^3 in which there are three component directions representing x_1, x_2, x_3; this is indeed a mapping of six dimensional solution space E^6 where all the components $x_1 \ldots x_6$ are represented.

Requirement Space E^3 in which each of the components of a_1, a_2, a_3, b may be represented.

The choice of three component directions in both the cases is deliberate and does not diminish the generality of the illustration. Introduce the slack variables $x_4, x_5, x_6 \geqslant 0$ and rewrite (A.1.3) as,

$$\begin{bmatrix} 2 & 1 & 1 & 1 & & \\ 1 & 2 & 1 & & 1 & \\ 3 & 1 & 1 & & & 1 \end{bmatrix} \cdot \begin{bmatrix} x_1 \\ x_2 \\ x_3 \\ x_4 \\ x_5 \\ x_6 \end{bmatrix} = \begin{bmatrix} 6 \\ 7 \\ 8 \end{bmatrix} \tag{A.1.4}$$

or

$$[a_1, a_2, a_3, e_1, e_2, e_3] \cdot \begin{bmatrix} x_1 \\ x_2 \\ x_3 \\ x_4 \\ x_5 \\ x_6 \end{bmatrix} = b$$

In the "solution space" figure A.1.1 any corner point where three planes meet obtained by putting corresponding non-basic variables to zero, represents a basic solution.

Example

Corner Point: P1: (0, 0, 0, 6, 7, 8), $x_1 = 0$, $x_2 = 0$, $x_3 = 0$.
Corner Point P2: (2, 3, —1, 0, 0, 0), $x_4 = 0, x_5 = 0$, $x_6 = 0$.

Indeed the scalar components of the vectors representing the corner

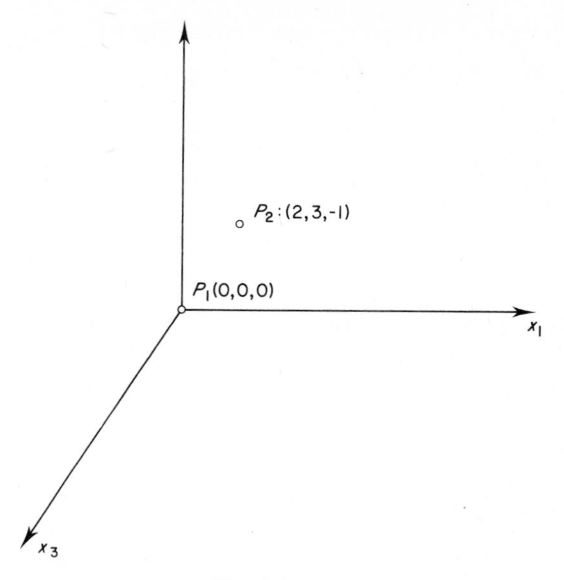

FIG. A.A.1.

points are the multipliers by which the basic column vectors are multiplied to obtain the solution vector. In the "requirement space" representation (Fig. A.1.2) one finds that any vector b in this space may be represented by a

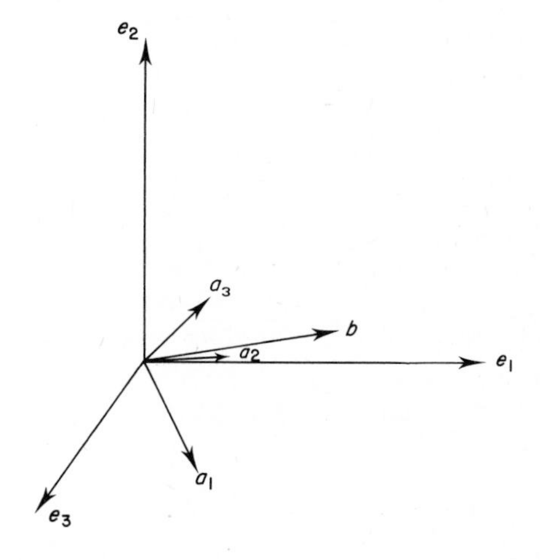

FIG A.1.2.

spanning set of vectors appropriately multiplied by scalars. The bases are linearly independent subset of these vectors. Maximum number of three make up the subset and each basis of necessity has only three vectors.

The point P1: This illustrates how b may be expressed as linear combination of e_1, e_2, e_3.
The point P2: This illustrates how b may be expressed as linear combination of a_1, a_2, a_3.

It can now be stated without proof that the columns of an nth order matrix B of rank n must be linearly independent. The matrix therefore must be non-singular and the columns (vectors) of the matrix must form a basis. For one may solve

$$Bx = b \tag{A.1.5}$$

with any n dimensional r.h.s. vector b and find a representation of b by linearly multiplying columns of B by the solution values. Conversely an nth order basis matrix B must have its columns linearly independent hence B must be non-singular.

We note that in linear programming only the non-negative multipliers of basis vectors which can obtain the solution vector is of interest. More precisely one is looking for basic feasible solutions in non-negative variables.

A.1.3 THEOREM OF FARKAS

A fundamental theorem concerning the solvability of a system of linear equation leads to further theorems which are of interest in mathematical programming. These latter theorems concern solvability of a system of linear equations in non-negative variables and the solvability of another set of equations made up of the transpose of the coefficient matrix of this system. The latter is known as Farkas's Theorem.

The systems $Ax = b$ has a solution if and only if the rank of A equals the rank of $[A:b]$; this is of course another way of stating [A1] the condition that b may be expressed as a linear combination of the columns of A (also see A.1.1). Hence the following theorem may be stated.

THEOREM. *For any $m \times n$ matrix A and any m-element vector b either*

$$Ax = b \tag{A.1.6}$$

has a solution x,

or
$$\begin{bmatrix} A' \\ b' \end{bmatrix} y = \begin{bmatrix} 0 \\ k \end{bmatrix} \tag{A.1.7}$$

has a solution y for any non zero k, but not both.

Proof

If rank $[A]$ = rank $[A \vdots b]$ then rank $\begin{bmatrix} A' \\ b' \end{bmatrix} <$ rank $\begin{bmatrix} A'0 \\ b'k \end{bmatrix}$, $k \neq 0$.

Thus if (A.1.6) has a solution then (A.1.7) has no solution. Conversely if

$$\text{rank}\begin{bmatrix} A' \\ b' \end{bmatrix} = \text{rank}\begin{bmatrix} A'0 \\ b'k \end{bmatrix} \text{ and } k \neq 0$$

then rank $[A] <$ rank $[A \vdots b]$.

Hence if (A.1.7) has a solution (A.1.6) has no solution.

A key feature of the above property is that

$$\text{rank } [A] + 1 = \text{rank}\begin{bmatrix} A'0 \\ b'k \end{bmatrix}, \qquad k \neq 0$$

under all circumstances.

This property may be illustrated geometrically for the following case. Let

$$A = [a_1, a_2] = \begin{bmatrix} a_{11} & a_{12} \\ a_{21} & a_{22} \\ a_{31} & a_{32} \end{bmatrix}, \qquad m = 3, n = 2, \tag{A.1.8}$$

and

$$b = \begin{bmatrix} b_1 \\ b_2 \\ b_3 \end{bmatrix}$$

be a column vector, then either no solution can be found for

$$Ax = b,$$

or there exists a vector y perpendicular to each column of A but not perpendicular to $b (k \neq 0)$.

If b does not lie on the plane P containing a_1, a_2 then $Ax \neq b$. In this case it is possible to find a vector $y = (y_1, y_2, y_3)'$ such that $A'y = 0$ but $b'y \neq 0$. Fig. A.1.3 illustrates this when y is a vector perpendicular to the plane P.

The theorem of separating hyperplanes may now be stated which is often used to deduce the duality properties and also the Kuhn–Tucker conditions.

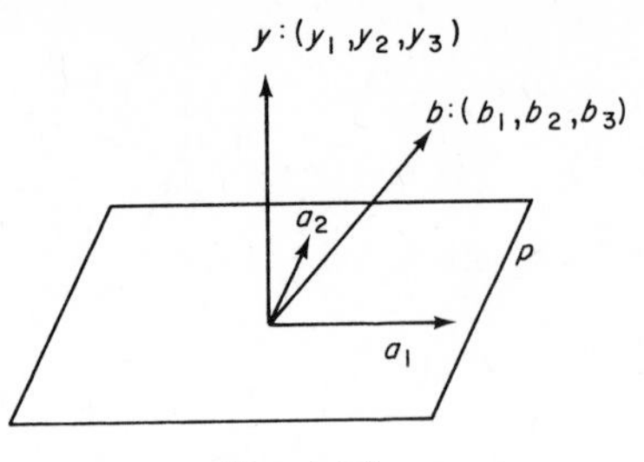

FIG. A.1.3.

THEOREM. *For any $m \times n$ matrix A and an m-component vector b, either*

$$Ax = b, \qquad x \geqslant 0 \tag{A.1.9}$$

has a solution x, (in non-negative components), or

$$A'y \geqslant 0, \qquad b'y < 0 \tag{A.1.10}$$

has a solution y, but not both.

The property stated by the theorem may be expanded as follows: either b is a vector contained in the cone spanned by the columns of A (such that $Ax = b$, $x \geqslant 0$) or there exists a vector y which makes non-obtuse angles with all the columns of A and an obtuse angle with b. The plane at right angles to the vector y is the separating hyperplane: it separates b from the cone spanned by the columns of A. Figures A.1.4, A.1.5, illustrate two

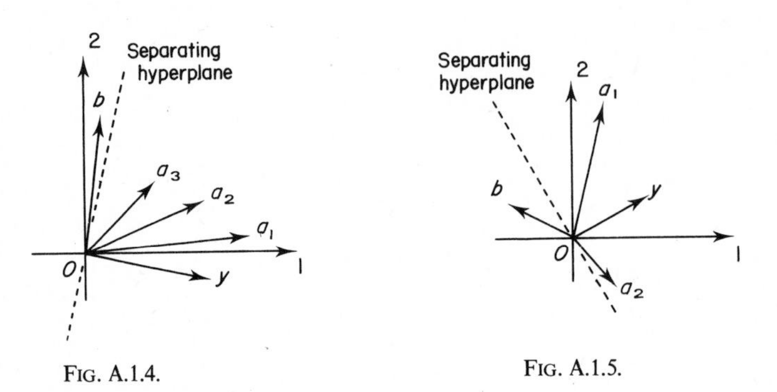

FIG. A.1.4. FIG. A.1.5.

situations where b is in the same plane as the columns of A but is not contained in this cone. It is easy to visualize the separating hyperplane also in the case $Ax \neq b$ for all x (positive and negative) when an illustration of the type Fig. A.1.3 need to be used.

Another statement of the same theorem is the Minkowski–Farkas Lemma.

LEMMA. *If $b'y \geqslant 0$ for all y such that $A'y \geqslant 0$ then $Ax = b$ is solvable in non-negative x.*

A.1.4 SOME PROPERTIES OF FUNCTIONS, LINEARITY, CONVEXITY, QUASI-CONVEXITY AND SOME PROPERTIES OF CONSTRAINT SETS

Even when an analyst is taking recourse to standard mathematical programming packages for the solution of optimization problems, it is important that he understands a few basic and hence far reaching properties of his model. This understanding is necessary to develop solution techniques for such models and also to appraise the solutions if they are obtained by some known techniques.

Linearity. A function $f(x)$ is said to be linear if given any two points x', x'' in the n-dimensional euclidean space E^n over which the function is defined, the equality

$$f(x'[1 - \lambda] + \lambda x'') = [1 - \lambda]f(x') + \lambda f(x'') \tag{A.1.11}$$

holds for all values of $0 < \lambda < 1$.

This is of course not true for integer linear forms, hence integer linear problems are nonlinear. In general a function for which the above relationship does not hold is said to be nonlinear.

Convexity. A function $f(x)$ defined over E^n is said to be convex if it is never underestimated by linear interpolation—a succinct definition due to Beale [2]. Mathematically this implies that a function $f(x)$ is convex if

$$f(x'[1 - \lambda] + \lambda x'') \leqslant f(x')[1 - \lambda] + \lambda f(x'') \tag{A.1.12}$$

$$\text{for all } x, x'' \in E^n \text{ and } 0 < \lambda < 1.$$

The concept of convexity also applies to sets. A set S is said to be convex if for any two points x' and x'' belonging to the set linear interpolate

$$z = [1 - \lambda]x' + \lambda x''; \qquad 0 < \lambda < 1 \tag{A.1.13}$$

also belongs to the set. That is, $z \in S$ for all x', $x'' \in S$. If a function $f(x)$ is convex then it is obvious that the constraint set generated by the inequality,

$$\{S: f(x) \leqslant b\} \tag{A.1.14}$$

is convex.

The function $f(x)$ is *strictly convex* if the less than equal sign in A.1.12 and

A.1.1.14 are replaced by a less than sign. A convex function admits of a global minimum and this is unique if the convexity is strict. For a twice differentiable function this is achieved where all the partial derivatives of the function vanishes.

A twice differentiable function $f(x)$ is convex or strictly convex if the quadratic form as associated with the Hessian matrix $\nabla^2 f(x)$ is positive semi-definite or definite respectively everywhere in the domain over which the function is defined.

If a function $f(x)$ is convex then the function $-f(x)$ is concave and vice versa. The only function that generates a convex set by equality constraint is the linear function. Some properties of the functions and constraints sets are tabulated in Table A.1.1.

It is perhaps important that one understands the properties shown in the above table: many a misconception and incorrect programming techniques have resulted due to a lack of appreciation of such properties.

Now consider a typical convex constraint set

$$x_1 + x_2 = 1 \tag{A.1.15}$$

if one squares both the sides of this equation the constraint set

$$x_1^2 + x_2^2 + 2x_1x_2 = 1 \tag{A.1.16}$$

is no longer convex.

Quasi convexity. This is a property weaker than convexity but is strong enough to have the majority of the desirable properties of mathematical programming which convexity possesses.

The function $f(x)$ is quasi-convex if given any two points x', x'' the relationship

$$f([1 - \lambda]x' + \lambda x'') \leqslant \max\{f(x'), f(x'')\}, \qquad x', x'' \in S \text{ and } 0 < \lambda < 1 \tag{A.1.17}$$

holds, when S is a set over which the function is defined. Consider the function $x^2/(1 + x^2)$ shown in Fig. A.1.6. The function is not convex simply because its values "do not rise fast enough" for large x. Often a nonlinear change of scale to the values ($y = f(x)$ of the original function) makes the new function convex. In this example the function

$$g(f(x)) = g(y) = 1/(1 - y) = 1 + x^2$$

is convex. A function is quasiconcave if its negative is quasiconvex.

TABLE A.1.1

Type	Function $\phi(x)$	Function $-\phi(x)$	Set S_1 $\phi(x) \leqslant b$	Set S_2 $\phi(x) \geqslant b$	Set $Se = S_1 \cap S_2$ $\phi(x) = b$	Example
Not Linear	Convex	Concave	Convex	Nonconvex	Nonconvex	$\phi(x_1, x_2) = a_1x_1^2 + a_2x_2^2$ $a_1, a_2 \geqslant 0$
	Nonconvex	Nonconvex	Nonconvex	Nonconvex	Nonconvex	$\phi(x_1) = a_1x_1^3 + a_2x_2 - a_3$ $a_1, a_2, a_3 \neq 0$
Linear	Convex	Convex	Convex	Convex	Convex	$\phi(x_1, x_2) = a_1x_1 + a_2x_2$

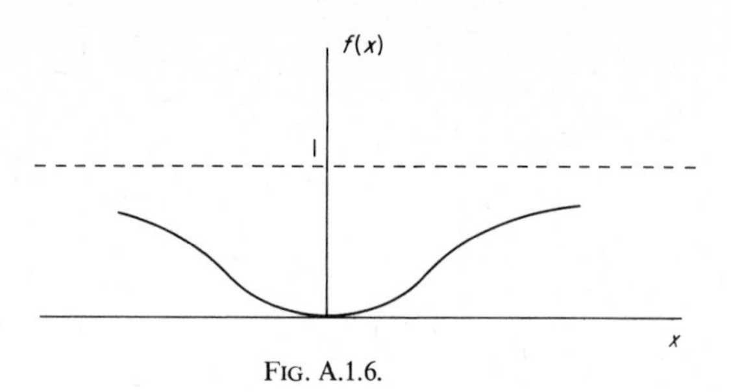

FIG. A.1.6.

A.1.5 DERIVATIVES OF FUNCTIONS OF MANY VARIABLES

Let the function $f(x_1, x_2 \ldots x_n)$ often expressed as $f(x)$ be such that its first and second partial derivatives exist; this is often expressed by the notation $f(x) \in c^1$ and $f(x) \in c^2$.

(a) *Gradient vector*

For a function $f(x) \in c^1$, the first partial derivative of the function with respect to each component direction defines a vector called the gradient vector at any point,

$$\nabla f\bar{x} = \left(\frac{\partial f}{\partial x_1}, \frac{\partial f}{\partial x_2}, \ldots \frac{\partial f}{\partial x_n}\right)\Big|_{\bar{x}}. \tag{A.1.18}$$

For the function

$$f(x_1, x_2, x_3) = a_1 x_1^2 + a_2 x_2^2 + a_3 x_3^2, \nabla f\bar{x} = (2a_1\bar{x}_1, 2a_2\bar{x}_2, 2a_3\bar{x}_3). \tag{A.1.19}$$

(b) *Hessian matrix*

Each of the partial derivatives of $f(x)$ may be differentiated again (provided $f \in c^2$) and these partial derivatives form a matrix of elements $\partial^2 f/(\partial x_i \partial x_j)$.

Such a square matrix is symmetric and is called the Hessian (H) of the function $f(x)$.

$$H = \begin{bmatrix} \frac{\partial^2 f}{\partial x_1^2}, \frac{\partial^2 f}{\partial x_1 \partial x_2}, \ldots, \frac{\partial^2 f}{\partial x_1 \partial x_n} \\ \frac{\partial^2 f}{\partial x_2 \partial x_1}, \frac{\partial^2 f}{\partial x_2^2}, \ldots, \frac{\partial^2 f}{\partial x_2 \partial x_n} \\ \frac{\partial^2 f}{\partial x_n \partial x_1}, \frac{\partial^2 f}{\partial x_n \partial x_2}, \ldots, \frac{\partial^2 f}{\partial x_n^2} \end{bmatrix} \tag{A.1.20}$$

For the function $f(x_1, x_2, x_3)$,

$$f(x_1, x_2, x_3) = c_1x_1 + c_2x_2 + c_3x_3 + \tfrac{1}{2}(x_1, x_2\, x_3)\cdot\begin{bmatrix} a_{11}a_{12}a_{13} \\ a_{21}a_{22}a_{23} \\ a_{31}a_{32}a_{33} \end{bmatrix}\cdot\begin{bmatrix} x_1 \\ x_2 \\ x_3 \end{bmatrix},$$

the Hessian matrix is

$$H = \begin{bmatrix} a_{11}a_{12}a_{13} \\ a_{21}a_{22}a_{23} \\ a_{31}a_{32}a_{33} \end{bmatrix}. \tag{A.1.21}$$

(c) *Jacobian matrix*

The m functions $g_i(x_1, x_2, \ldots, x_n)$; $i = 1, 2, \ldots, m$, when differentiated with respect to each of the component directions $x_1, x_2, \ldots, x_n$ lead to m vectors of n components which may be set out in the following matrix form;

$$\begin{bmatrix} \frac{\partial g_1}{\partial x_1}, \frac{\partial g_1}{\partial x_2}, \ldots, \frac{\partial g_1}{\partial x_n} \\ \frac{\partial g_2}{\partial x_1}, \frac{\partial g_2}{\partial x_2}, \ldots, \frac{\partial g_2}{\partial x_n} \\ \vdots \quad \vdots \quad \quad \vdots \\ \frac{\partial g_m}{\partial x_1}, \frac{\partial g_m}{\partial x_2}, \ldots, \frac{\partial g_m}{\partial x_n} \end{bmatrix}. \tag{A.1.22}$$

Any square submatrix of dimension $m \times m$ of the above $m \times n$ matrix is called the Jacobian Matrix of the variable with respect to which the functions are differentiated i.e.,

$$J\bar{x}(1, 2, \ldots m) = \begin{bmatrix} \frac{\partial g_1}{\partial x_1}, \frac{\partial g_1}{\partial x_2}, \ldots, \frac{\partial g_1}{\partial x_m} \\ \vdots \\ \frac{\partial g_m}{\partial x_1}, \frac{\partial g_m}{\partial x_2}, \ldots, \frac{\partial g_m}{\partial x_m} \end{bmatrix}_{\bar{x}} \tag{A.1.23}$$

A.1.6 IMPLICIT FUNCTION THEOREM

For a set of constraints,

$$g_i(x_1, x_2, \ldots, x_n) = b_i, \qquad i = 1, 2, \ldots, m, \tag{A1.24}$$

the implicit function theorem states that if there exists a point x^* in E^n such that,

(i) the function $g_i(x) \in c^1$, $i = 1, 2, \ldots, m$ in some δ neighbourhood of x^*,

(ii) $g_i(x) = b$, $i = 1, 2, \ldots, m$, and (A.1.25)

(iii) the Jacobian $Jx^*(j_1, j_2, \ldots, j_m)$ is nonsingular

then there exists an epsilon-neighbourhood (epsilon > 0) of

$$\hat{x} = [x^*_{k_1}\ x^*_{k_2}, \ldots, x^*_{k_{n-m}}]$$

in E^{n-m} and that for every point $\hat{x}$ in this neighbourhood there exists a set of functions single valued and continuous for which the variables in the Jacobian may be expressed as

(a) $$x^*_{j_i} = \phi_i(\hat{x}), \qquad i = 1, 2, \ldots, m, \qquad \text{(A.1.26)}$$

(b) the x^* so evaluated satisfies (A.1.5.7)

(c) the functions $\phi_i(x)$ are differentiable and for the directions k_r, $r = 1, 2, \ldots, n - m$ these derivatives may be obtained by solving the set of equations

$$\sum_{u=1}^{m} \frac{\partial g_i}{\partial x_{j_u}} \cdot \frac{\partial \phi_u}{\partial x_{k_r}} = \frac{\partial g_i}{\partial x_{k_r}}, \qquad i = 1, 2, \ldots, m \qquad \text{(A.1.27)}$$

The implicit function theorem may be invoked to deduce the Lagrange Multiplier relationships.

REFERENCE

A1. Hadley, G., "Linear Algebra", Addison-Wesley, Reading, Mass. 1961.

APPENDIX 2

On Using a Linear Programming System

A.2.1 INTRODUCTION

In order to be able to use a mathematical programming system some idea of the computational methods for solving mathematical programs is necessary. It is just as important and to the point, that the general terminology used by the practitioners should be understood; this appendix is meant to explain some of the terminology in vogue and in general facilities which one might expect to find in a mathematical programming system.

Section A.2.2 describes the typical and perhaps most often used input format for data: MPS360 data format. Section A.2.3 outlines some features of control languages, a list of typical verbs/procedures/subroutines and a list of controlling communication variables is provided and their functions explained. An example illustrating the presentation and solution of a small problem using XDLA is described in Section A.2.4. Sections A.2.3 and A.2.4 are based on the facilities available in XDLA mathematical programming system for the ICL 1900 series computers.

A.2.2 INPUT FORMAT (MPS 360)

A problem is specified usually in terms of BCD card images, such representation being actually on cards or on magnetic tape or on disk files. The systems procedures such as INPUT, REVISE etc., process these information. A set of card images specifying a problem is often referred to as the problem file.

In all cases, there are two types of cards in the problem file:

1. Indicator cards which specify the type of data which is to follow.
2. Data cards which contain the actual data values.

An indicator card is made up of alphabetic characters starting always in character position 1, it must be one of the types defined in the rest of this section.

A data card follows the same fixed format made up of fields defined below. The contents of the fields may have different significance depending upon the section of data in which these appear.

	Field 1	Field 2	Field 3	Field 4	Field 5	Field 6
Columns	2–3	5–12	15–22	25–36	40–47	50–61
Contents	Code	Name	Name	Value	Name	Value

Name character set

This is made up of the characters A through Z and 0 to 9. A name made up of name character set must start with an alphabetic character.

The names and codes are made up of alphanumeric characters and the names must be left justified, no embedded blanks are allowed. The value column will accept anything that may be presented as F12.p and Iq formats for FORTRAN where $p < 12$ and $1 < q < 12$. A card with asterisk (*) in the first column is not input and treated as comment card.

In a data card a "£" sign as the first character in field 3 or 5 indicates that the information from that point until column 71 is treated as comment.

The data deck

The data deck is made up of the following set of indicator and data cards. The NAME card gives a user specified name to the problem: the name is of course made up from the name character set.

Columns	1–4	15–22
Contents	NAME	User given name

The ENDATA card simply announces the end of the current data deck i.e. the problem specification started by a NAME card.

Columns	1–6
Contents	ENDATA

Rows

Rows indicator cards follow the NAME card and announces that the data cards to follow will specify row (constraint) names.

Columns	1–4
Contents	ROWS

Rows data cards

Field 1: Defines constraint type.

∧N or N∧	Free row, i.e., no constraint. May be used as objective or change row.
∧G or G∧	Greater than or equal to
∧L or L∧	Less than or equal to
∧E or E∧	Implies equality
Dx	where x may be N, G, L, E this means that the coefficients of this row will be generated as a linear combination of other rows.

For further details concerning Dx code and other rows facilities see [A2].

Field 2: This simply defines the Rows name; the contents are of course made up of the name character set.

Columns

Column cards specify the names to be assigned to the columns (structural variables) in the LP matrix, and define, in terms of column vectors, the actual values of the matrix elements.

Columns	1–7
Contents	COLUMNS

Data Cards

Field 1: blank (ignored)

Field 2: gives the name of the column that is to contain the elements specified in the following fields. The name is made up of the name character set.

Field 3: contains the name of a row.

Field 4: used in conjunction with Field 3 contains the value of the matrix element.

Field 5: Is optional may be used like field 3.

Field 6: Is optional may be used like field 4.

Note that only non-zero matrix elements need to be specified in the COLUMNS section. If not explicitly specified otherwise the matrix coefficient is taken to be zero.

RHS

RHS cards specify the name of the RHS vector of the problem or the change rhs vector which may be used for parametric programming; the numerical values of the components of the vector are also defined by the data cards.

Columns	1–3
Contents	RHS

The data cards which follow have exactly the same format as columns cards; except the column name in this case signifies the RHS name.

BOUNDS

Bounds cards specify limits on the values of the structural variables. If the variable is not specified in the bound set then it is automatically assumed to lie between 0 and $+\alpha$. Like a RHS vector which is given a name, the set of variables in one bound set is also given a name.

Columns	1–6
Contents	BOUNDS

Data cards

Field 1: Specifies the type of bound:

LO Lower bound
UP Upper bound
FX Fixed value of the variable
FR Free variable ($-\alpha$ to $+\alpha$)

MI Lower bound is $-\alpha$
PL Upper bound is $+\alpha$

Field 2: Identifies a name for the row or bound set.
Field 3: Identifies the column name of the variable belonging to this set,
Field 4: Identifies the value of the bound; this has a numeric value only in association with LO, UP, FX in Field 1, otherwise it is blank,
Field 5: Is blank and ignored,
Field 6: Is blank and ignored.

Detailed information concerning further facilities and also the format of REVISE data cards are to be found in [1].

A.2.3 SYSTEMS CONTROL LANGUAGE AND SYSTEMS PROCEDURES

Different systems employ different control languages, which are some way akin to one of the high level languages such as FORTRAN or ALGOL. The systems control language used in XDLA ICL 1900 series machine is similar to ALGOL but with very limited facilities: a full description of the language is to be found in [A3].

For all practical purposes, however, a user may input his problem, apply suitable algorithms, and output or monitor his solution knowing a few standard systems procedures. Some further control on the processes may be achieved by setting Linking variables or CR cells which allow interaction with the systems procedures. The more important systems procedures and linking variables are briefly described in this section.

Input/monitor procedures

INPUT ($S(g)$, t)

This procedure reads from the currently defined source file [A3] the problem matrix presented in BCD card images. The linking variable £LPROBN will contain the problem name. The parameter t is a string specifying alternative input formats. In the present instance this should be "MPS/360".

LIST

This procedure provides a list of information concerning the current problem as presented in the source file. Information listed is:

The number of rows and columns of each type

All row and column names together with the number of non-zero elements in each row and column

Some statistics concerning the minimum and maximum of these elements and their mean etc.

PICTURE

This procedure prints out using coded characters the matrix elements and the structure of the problem matrix. The layout and the legends such as,

$$* \text{ if } a_{ij} = 1$$

$$1 \text{ if } 1 < |a_{ij}| < 10, \text{ etc.,}$$

are explained in the monitor information from the procedure.

READBASIS

This procedure inputs a starting basis matrix presented in BCD card images: a call to INVERT should be made to obtain the inverse of this basis and carry on computation.

Algorithmic procedures

INVERT

This procedure carries out the inversion of the currently defined basis matrix. Some information concerning the original and inverted matrices and the structure of the matrix are also monitored.

CRASH

This is a procedure meant to create a starting basis quickly so that this may reduce the sum and number of infeasibilities or improve the value of the objective function.

PRIMAL

This is the most important procedure of linear programming; given the problem and a starting basis this procedure carries out the optimisation steps by the revised simplex method. The problem to be solved is defined by:

£LRHSO text variable containing the name of the RHS vector, +£LCMULT1*£LRHS1+£LCMULT2+£LRHS2, these latter terms are added only if a composite RHS is used.

£LOBJO text variable which contains the name of the objective row, + £LRMULT1*£LOBJ1 + £LRMULT2*LOBJ2

These latter terms are added only if a composite objective row is used.

The linking variable £LREAS contains an integer value signifying the reason for termination.

£LREAS = 0 indicates optimal feasible solution
= 1 indicates no feasible solution
= 2 indicates unbounded solution
= 20 indicates iteration limit reached etc.,

The iteration report from this procedure is more or less self explanatory and contains information concerning:

Iteration number, incoming and outgoing variable, the reduced cost coefficient, value of the objective function, the sum of infeasibilities.

RHSPARA

This procedure performs parametric programming on the right hand side £LRHS0 + £LCMULT1 * £LRHS1 + £LCMULT2 + £LRHS2.

The value of £LCMULT2 is changed in discrete steps for each basis change. Its value is increased if £LPMAX is positive or decreased if £LPMAX is negative. The rest of the problem definition is as for PRIMAL. The iteration report is very similar to that of PRIMAL. The basis changes and change in £LCMULT2 continues until:

The maximum (or minimum) value of £LPMAX is reached.
The parameter £LCMULT2 becomes unbounded.
The problem goes nofeasible for any further increase in the value of £LCMULT2.

OBJPARA

This procedure performs parametric programming on the objective row £LOBJ0 + £LRMULT1*£LOBJ1 + £LRMULT2*£LOBJ2. The value of £LCMULT2 is changed in discrete steps and is increased if £LPMAX is positive and decreased if £LPMAX is negative. The rest of the problem definition is as for PRIMAL. The basis changes and the change in £LCMULT2 continues until:

The maximum (or minimum) value of £LPMAX is reached.

The parameter £LCMULT2 becomes unbounded.
The solution goes unbounded.

TABLEAU (i, j, k)

This procedure produces the updated values of the transformed coefficients corresponding to each column in the problem. If j is zero or omitted then the stack variables and the basic columns are omitted. If $j = 1$ the basic columns are omitted ... if $j = 3$ the full tableau is output.

INTEGER

This procedure solves a pure or a mixed integer linear programming problem employing a branch and bound method. The method of preparing input to such a problem is described in [A3]. It is indeed possible to have sophisticated control on the solution process; this is also considered in detail in this manual.

Output procedure

SOLUTION

This procedure prints out the solution value and some associated information using the currently available basis for the problem under consideration: the problem of course is the one defined by the PRIMAL procedure. The output is naturally divided into ROW and COLUMN sections. In the ROW INFORMATION, the constraint name, row type, solution value, RHS value and the shadow price are printed. In COLUMN INFORMATION, variable name, type, solution value, bounds, cost coefficient and reduced cost coefficient are printed.

PUNCH BASIS

This procedure outputs the current basis in external BCD format on a specified peripheral. The output on card, disk or tape may be subsequently read by READ BASIS procedure.

A full list of the linking variables, their function and their standard values are described in [A3].

A.2.4 SETTING UP AND RUNNING A PROBLEM

To set up and run a problem it is necessary to call some systems procedures to define files to be used; these are in addition to I/O and Algorithmic pro-

	SOURCE 11	SOURCE 12	SOURCE 21	SOURCE 22	SOURCE 31	SOURCE 32		B1
PROFIT	2·0	1·5	1·0	2·5	2·0	2·5		
ROW 1	3·0		0·5		3·0		≤	8·0
ROW 2		2·0		0·5		0·5	≤	4·0
DEMAND 1	1·0	1·0					=	2·0
DEMAND 2			1·0	1·0			=	5·0
DEMAND 3					1·0	1·0	=	4·0

```
NAME              LP MODEL
ROWS
 N   PROFIT
 L   ROW1
 L   ROW2
 E   DEMAND1
 E   DEMAND2
 E   DEMAND3
COLUMNS
     SOURCE11   PROFIT     2·0
     SOURCE11   ROW1       3·0
     SOURCE11   DEMAND1    1·0
     SOURCE12   PROFIT     1·5
     SOURCE12   ROW2       2·0
     SOURCE12   DEMAND1    1·0
     SOURCE21   PROFIT     1·0
     SOURCE21   ROW1       0·5
     SOURCE21   DEMAND2    1·0
     SOURCE22   PROFIT     2·5
     SOURCE22   ROW2       0·5
     SOURCE22   DEMAND2    1·0
     SOURCE31   PROFIT     2·0
     SOURCE31   ROW1       3·0
     SOURCE31   DEMAND3    1·0
     SOURCE32   PROFIT     0·5
     SOURCE32   ROW2       0·5
     SOURCE32   DEMAND3    1·0
RHS
     B1         ROW1       8·0
     B1         ROW2       4·0
     B1         DEMAND1    2·0
     B1         DEMAND2    5·0
     B1         DEMAND3    4·0
ENDATA
```

FIG A.2.1. Weighted Distribution Model in MPS/360 Input Format.

cedures. The following control program is illustrative of various features of the system.

```
'BEGIN'
INPUT ("LPMODEL", "MPS/360")
PICTURE
LIST(1)
'COMMENT'...THE MATRIX CALLED LPMODEL IS INPUT
'COMMENT'...ASSUME 1000 AS A LIMIT OF ITERATIONS
£LITLIM:=1
CONT:'FOR' £WI2:=1 'STEP' 1 'UNTIL' 1000 'DO'
'BEGIN'
£LPH1:=0.0
£LMAJI:=1
PRIMAL
TABLEAU (,3,1)
£LPITEM:=1
SOLUTION
£LPITEM:=15
'IF' £LREAS 'EQ' 0 'THEN' 'GO TO' OPTIMUM
'IF' £LREAS 'EQ' 1 'THEN' 'GO TO' NOFEAS
'IF' £LREAS 'EQ' 2 'THEN' 'GO TO' UNBOUND
'END'
'GO TO' NEXT
OPTIMUM:SOLUTION
NOFEAS: 'GO TO' NEXT
UNBOUND: 'GO TO' NEXT
NEXT: 'END'
'FINISH'
```

A problem set up in MPS/360 format (see Figure A.2.1) and driven by the above control program will input the problem, print a picture and a list. It will then be processed with tableau and solution monitored at each iteration. If OPTIMUM is reached a complete solution print is obtained.

REFERENCES

A.2. Mathematical Programming System Extended (MPSX)...Program Description, PROGRAM NUMBER 5734-XM4 IBM Technical Publications Dept., White Plains, N.Y. 10601.

A.3. XDLA–MARK-3 LPSYSTEM. Technical Publication 4147 International Computers Ltd., CIL House, Putney, London, S.W.15.

APPENDIX 3

UIMP: User Interface to Mathematical Programming a Modelling Language

A.3.1. INTRODUCTION

A number of Matrix Generator Report Writer (MGRW) systems are in use in industry and to any organization which uses Mathematical Programming (MP) as a serious modelling tool such systems are of considerable value. Without exception such systems are used in conjunction with proven MP software: [A4] contains a brief survey of such systems.

To understand the requirements of an MGRW system in which Matrix Generator (MG) programs and Report Writer (RW) programs are written it is necessary to identify the tasks performed by an MG or a RW. An MG or a RW may of course be written in a high level computer language such as FORTRAN, ALGOL, PL1, or equally in an MGRW language. A typical modelling system using an MG, an OPTIMIZER, and a RW as set out in Fig. A.3.1 may work in the following way. The PROBLEM DATA is pre-

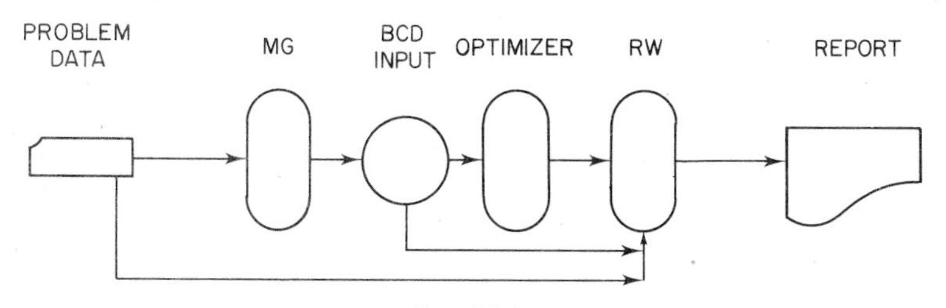

FIG. A.3.1.

sented in the form of sets of tabulated information. These are read and processed by an MG program which produces a BCD INPUT FILE. This usually contains,

(a) the logical (i.e. slack and artificial) variables names or row names which are coded by suitable text expressions.

(b) the structural variables names or column names which are similarly coded,
(c) the coefficients of the problem matrix,
(d) the Right Hand Side (RHS), BOUNDS and RANGES information,
(e) some information concerning the starting basis.

A RW is used primarily to extract only the pertinent information from the solution obtained by the OPTIMIZER, and it presents this information in a suitably tabulated format.

The RW usually consults the PROBLEM DATA held as tables, and the BCD INPUT; it may also carry out some arithmetical operations on these solution values. An MGRW system therefore incorporates at least the following features

(i) input of problem data in tabular form,
(ii) construction of row and column names by name expressions,
(iii) using constants or arithmetic expressions to specify the matrix, RHS, BOUND etc coefficients by suitable row or column generator clause (procedure),
(iv) accessing the solution file to obtain solution values, reduced costs, ranges etc.,
(v) format and print tabular information.

These are of course the features of all the known conventional MGRW systems. The UIMP system in addition provides a great emphasis on the data structuring aspect of MP modelling. To illustrate this the following observations may be made. One important aspect of real life MP models is that the variables as well as constraints of these models usually appear in groups. Such groupings may be typically found in models involving multi-time-periods, multi-products, or with decentralized facilities etc. This type of structure of a model taken from real life is naturally reflected in the input data structure and the structure of the constraint matrix. In the UIMP language a data structure facility is provided explicitly whereby such models in spite of their apparent complexity may be constructed in a straight forward fashion. In Section 2 an overview of the language is presented and the syntactical components are defined. In Section 3 a small but realistic problem is presented. The problem is illustrative of a structure commonly found in practice; the tactics adopted to formulate this model explain the principle of use of the language. In Section 4 some ramifications of the data structure facility are presented. In Section 5 the illustrative problem is fully coded in UIMP language and the various steps have been annotated.

A.3.2. AN OVERVIEW OF THE LANGUAGE AND THE BASIC SYNTACTIC COMPONENTS

The UIMP system is essentially a high level language compiler and executor; the language is of course limited in its generality. The language provides facilities for easy definition and manipulation of tabular data. In addition to numeric or text information one may also define structure made out of structure elements. The structure elements can be used in name expressions to construct names for the ROWS, COLUMNS, RHS, BOUNDS etc for an LP model. As a part of the language a few generator clauses are provided which are used to construct the model by "rows" or by "columns". These clauses require (cf syntax later) that name expressions be provided for model row and column names and arithmetical expressions be used to specify the model coefficients. The language possesses a block structure i.e., a set of UIMP statements may be bracketed by BEGIN and END commands. The language also incorporates table manipulative and table print facilities which find use in RW programs.

The basic syntax of the language is described using the Backus Naur Form (BNF), cf [A5]. This brief description is supplemented by explanatory notes where appropriate. The restricted size of the present appendix imposes some incompleteness in the description of the language however, a detailed account can be found in [A6].

A.3.2a. Preliminary syntax

⟨basic symbol⟩: : = ⟨letter⟩|⟨digit⟩|⟨delimiter⟩

where : : = is the metalanguage connective meaning 'is syntactically defined as', and | is the other connective separating alternative definitions and may be read as 'or'. The symbols ⟨ ⟩ are metalanguage brackets used to enclose any subcomponent of the definition which may itself contain the 'OR' symbol | and brackted subcomponents nested to any depth. For example

$$\langle a\rangle::=\langle b\rangle\langle c\rangle|\langle d\rangle\langle|\langle e\rangle\ |\langle f\rangle\rangle$$

is read as 'a is defined as either b followed by c, or d followed optionally by e or f'; therefore ⟨|...⟩ indicates an optional subcomponent that may be omitted. The metalanguage definition may now be continued as set out below,

⟨letter⟩: : = $A|B|\ldots|Y|Z$
⟨digit⟩: : = 0|1|2...|9
⟨integer⟩: : = ⟨−⟩⟨digit⟩|⟨digit⟩|⟨integer⟩⟨digit⟩
⟨arithmetic operator⟩: : = + | − |*|/

⟨relational operator⟩: : = LT|LE|EQ|GE|GT
⟨operator⟩: : = ⟨arithmetic operator⟩|⟨relational operator⟩
|⟨logical operators⟩|⟨sequential operator⟩|. . .
⟨delimiter⟩: : = ⟨operator⟩|⟨separator⟩|⟨bracket⟩|. . .
⟨logical operator⟩: : = AND|OR|NOT|XOR
⟨sequential operator⟩: : = GOTO|IF|THEN|FOR|DO|LET
† ⟨separator⟩: : = ,|$\wedge$|η|;|: = |STEP|UNTIL|IN|EXCEPT|BE
⟨bracket⟩: : = (|)|[|]|"|BEGIN|END
⟨identifier⟩: : = ⟨letter⟩|⟨identifier⟩⟨letter⟩|⟨identifier⟩⟨digit⟩
. . . . . .

Using these basic elements some typical statements and clauses of the language may be defined. There are in general four different modes of operation in the language these are

Integer, operations involving integer identifiers,
Real, operations involving real identifiers,
Text, operations involving text identifiers,
Structure, operations involving structure identifiers.

The character strings which identify the structure elements are themselves useful for name generation in the LP model. Hence these are made available for manipulation by the following syntactic relationship

⟨text string⟩: : = ⟨letter⟩|⟨digit⟩|⟨text string⟩⟨letter⟩|⟨text string⟩⟨digit⟩
⟨text⟩: : = "⟨text string⟩"|(⟨structure element⟩)
⟨name expression⟩: : = ⟨text⟩|⟨name expression⟩&⟨text⟩

A.3.2b. Simple declarator clauses

There are the following types of declarator clause

LET ⟨identifier⟩ BE INTEGER
LET ⟨identifier⟩ BE REAL
LET ⟨identifier⟩ BE TEXT
LET ⟨identifier⟩ BE STRUCTURE VARIABLE
LET ⟨structure identifier⟩ BE ⟨structure element list⟩

where

⟨structure element list⟩: : = ⟨structure element⟩⟨,|$\wedge$⟩
⟨structure element list⟩|⟨a list of structure elements defined in a FOR loop clause⟩

† $\wedge$ is the space character and η is the new line character.

This declaration clause of structure identifier and structure elements also defines these quantities. Note that these may be defined implicitly at the data entry stage by formatted tables, for further details see Section 4.

Examples

LET Q1 BE SA, SB, SC
LET SA BE P1, P2, P3
†LET THEAD BE FOR J: = 1 STEP 1 UNTIL 4 'TAIL'&CHAR(J)

These lead to the following two structures.

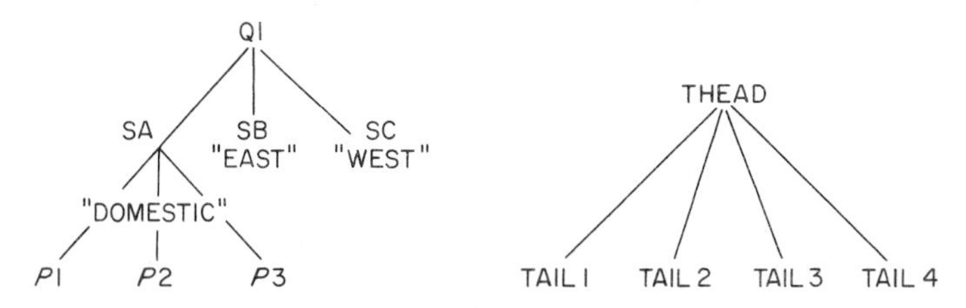

FIG. A.3.2.

In these two structures THEAD, Q1, SA are structure identifiers; SA, SB, SC, P1, P2, P3, TAIL1, . . . , TAIL4 are structure elements. Depending on the context therefore SA may be referred to (and used) as a structure identifier, implying there is a substructure made up of structure elements. It may also be referred to as an element which implies there is a superstructure in this case Q1 to which it belongs. The following properties are associated with structure elements and structure identifiers:

(i) A structure identifier referring to the head of the structure containing a set of structure elements has a cardinality which is the number of elements in the immediate substructure. This is expressed in the form N:⟨structure identifier⟩. Thus N:THEAD has the value 4.
(ii) A structure element has an ordinality within its superstructure expressed as I:⟨structure element⟩IN⟨structure identifier⟩ .Thus I:SB IN Q1 has the value 2. Also note that one may denote structure element SB by Q1(2).
(iii) The character string making up the name of the structure element is obtained by bracketing the structure element.

† CHAR(J) is a function which obtains the character representation of the integer parameter J.

(iv) One may associate with a structure element or a structure identifier text strings of any length. These may be referred to using the prefix T:. The following text assignment statements illustrate its use.

T:SA: = "DOMESTIC"

T:SB: = "EAST"

T:SC: = "WEST"

These can be used as text strings in the text expressions and the print statements etc.

A.3.2c. Declaration of tables

Tables are declared by the following declarator clause

LET TABLE⟨table identifier⟩BE⟨structure identifier⟩
DOWN BY⟨structure identifier⟩ACROSS⟨|TYPE⟨INTEGER
|REAL|TEXT⟩⟩

Note that the syntactic components TABLE, BY, DOWN, ACROSS etc do not appear in the short BNF definition provided earlier in the section 2a, however, here the context of their use makes these quite clear.

Example

LET TABLE EXAMPLE BE THEAD
DOWN BY Q1 ACROSS TYPE REAL

This leads to the table called EXAMPLE shown in the Fig. A.3.3 below

		Q1				
		SA			SB	SC
	EXAMPLE	P1	P2	P3		
THEAD	TAIL1					
	TAIL2					
	TAIL3					
	TAIL4					

Fig. A.3.3.

Like the structure definition the table definition can be made implicit at the time of data entry, cf. Section 4.

A.3.2d. Structure element context and enumeration

To identify a particular cell in a table such as the one shown in Fig. A.3.3

above requires that there should be a scheme to reference the components of the row and column structure. For example column 3 is identified by "P3 IN SA IN Q1" which is called an "element context expression" since each element is quoted in the context of its immediate superior in the hierarchy. An element may have many different superiors in the whole structure but a context does specify one particular superior at that moment. The syntax is given by the following definitions

⟨element reference⟩: : =
    ⟨structure element⟩|
    ⟨structure variable⟩|
    ⟨element reference⟩ (⟨integer expression⟩)

⟨element context expression⟩: : =
    ⟨element reference⟩|
    ⟨element reference⟩IN⟨element context expression⟩

Context is an important attribute of the structure variables and may be assigned with an element context expression. For example after carrying out the assignment statement

Q: = P3 IN SA IN Q1

one may use Q as a reference to the column in the table of diagram 3. Thus ... EXAMPLE(TAIL4 IN THEAD, Q) ... indicates position "row 4 by column 3".

Finally element context expressions may be used to enumerate repetitions over a set. This can be done by FOR and SUM clauses as follows

FOR ⟨element context expression⟩ DO ...

SUM ⟨element context expression⟩ ...

"FOR ... DO" enumerates repetition of a statement which follows "DO" (or a block of statements using BEGIN ... END) while "SUM" enumerates summation of a sub-expression within a linear form, cf. below section 2e. In each case the enumeration consists of allowing one or more structure variables to take all possible assignments indicated by the element context expression. One does not of course change any structure variables that already have an assignment prior to entering the loop. Hence a clear distinction exists between "assigned" and "de-assigned" structure variables and various rules apply, not stated here, which correspond to similar rules in other languages. Generally a structure variable assigned before entering a loop remains assigned after leaving the loop while any structure variable de-assigned before remains de-assigned again after.

A.3.2e. Linear and column form

The Row Generator and Column Generator clause defined in the next section uses the Linear Form and Column Form defined in this section.

```
⟨linear form⟩::=⟨linear term⟩|⟨linear form⟩⟨PLUS|MINUS⟩
  ⟨linear term⟩
⟨column form⟩::=⟨column term⟩|⟨column form⟩AND⟨column term⟩
⟨linear term⟩::=⟨arithmetic expression⟩MPLYING⟨name expression⟩
                |⟨SUM loop clause⟩⟨linear term⟩
                |⟨SUM loop clause⟩[⟨linear form⟩]
⟨column term⟩::=⟨arithmetic expression⟩ON⟨name expression⟩
                |⟨FOR loop clause⟩⟨column term⟩
                |⟨FOR loop clause⟩BEGIN⟨column form⟩END
```

A.3.2f. Generating the model

Rows and Columns of an LP model may be declared by statements in the following form

```
LET ⟨identifier⟩ BE VARIABLE
    ⟨|TYPE⟨PLUS|MINUS|FIXED|FREE⟩⟩
    ⟨|CONSTRUCTED ⟨element context expression⟩⟩
                ⟨|BY ⟨element context expression⟩⟩⟩
    CODE ⟨name expression⟩
```

Appearance of a "CONSTRUCTED" clause indicates a class of variables which may subsequently be referenced in essentially the same way as positions in a table illustrated in 3.2d above. Thus the same mechanisms are used to specify an individual variable and the same rules apply. Declaration of rows in an LP model is exactly similar and need not be given here separately.

This declaration does not actually generate a row or a column in the model. This is done by a row-generator statement taking the following form

```
    ⟨row-generator statement⟩::=ROW⟨constraint reference⟩
⟨|TYPE⟨LE|GE|EQ|FREE⟩⟩
⟨|IS⟨linear form⟩⟩
⟨|RHS⟨name expression⟩VALUE⟨arithmetic expression⟩⟩
⟨|RANGE⟨name expression⟩VALUE⟨arithmetic expression⟩⟩
⟨|IN BASIS⟨name expression⟩VARIABLE⟨variable reference⟩
            PIVOTS OUT ROW⟨|AT⟨LB|UB⟩⟩⟩
```

A column generator statement is used to generate a column of the model and is defined as

```
⟨column-generator statement⟩: : = COLUMN⟨variable reference⟩
    ⟨|TYPE⟨PLUS|MINUS|FIXED|FREE⟩⟩
    ⟨|HAS⟨column form⟩⟩
    ⟨|BOUNDS⟨name expression⟩
                ⟨|UP VALUE⟨arithmetic expression⟩⟩
                ⟨|LO VALUE⟨arithmetic expression⟩⟩
                ⟨|FX VALUE⟨arithmetic expression⟩⟩
    ⟨|IN BASIS⟨name expression⟩IS SET TO⟨UB|LB⟩⟩
```

Thus the column-generator is almost symmetric with the row-generator statements. The "IN BASIS" clause is unique to the UIMP system which allows "starting points", i.e. basis specification, to be supplied in an MG program, no other MGRW system to our knowledge provides this facility.

A.3.2g. Table print clause

The table print clause of UIMP makes RW programs easy and effective, it has the syntax

```
PRINTABLE⟨table identifier⟩
  ⟨|DOWN⟨structure identifier⟩⟨|,⟨format identifier⟩⟩⟩
  ⟨|ACROSS structure identifier⟩⟨|,⟨format identifier⟩⟩⟩
  ⟨|HEADER⟨format identifier⟩⟩
  ⟨|TRAILER⟨format identifier⟩⟩
```

An example of the use of this clause is to be found in Section 5 which contains a short RW program.

A.3.2h. Other syntactic features

There is a collection of numerous other features which have not been described. In this subsection some of the essential ones are briefly mentioned. The loop clauses may all include modifiers introduced by EXCEPT and a logical expression. Special table manipulative clauses are incorporated which allow

adding, multiplying, taking the ratio of two rows of one or more tables,
the above operations for the columns.

In order to access the information output by the Optimizer one needs only to reference solution values, values of reduced cost coefficients etc. These may be obtained by using the variables X:⟨name expression⟩, DJ:⟨name expression⟩. The example illustrates the use of some of these.

TABLE A.3.1. Table of machine-hours (TABH)

	SUMMER PERIOD (H1)								WINTER PERIOD (H2)							
	Normal (N) working hours			Over-time (O)			Total hours available (AV)		Normal (N) working hours			Over-time (O)			Total hours available (AV)	
	P1	P2	P3	P1	P2	P3	Normal W-hrs	Over-time	P1	P2	P3	P1	P2	P3	Normal W-hrs	Over-time
Machine 1 (M1)	4	5	6	3	4	6	100	80	5	6	7	4	5	5	110	90
Machine 2 (M2)	7	6	6	6	5	5	100	90	8	7	7	7	6	6	110	100
Machine 3 (M3)	3	–	–	2	–	–	40	30	4	–	–	3	–	–	50	40

P1 = Nuts P2 = Bolts P3 = Washers

A.3.3 PRINCIPLE OF USE AND MODELLING TACTICS ILLUSTRATED BY AN EXAMPLE

In this section an illustrative problem is first described and its linear programming formulation in mathematical notation is provided. A user formulation is then prepared. This is of course in a format which may be directly used for solution by an optimizer.

A.3.3a. An illustrative example

A company manufactures three products P1, P2, P3 (NUTS, BOLTS, WASHERS) and has at its disposal three machines M1, M2, M3. The company can undertake normal and overtime production and needs to plan for

TABLE A.3.2. Table of production costs (TABC)

	SUMMER PERIOD						WINTER PERIOD					
	Normal Working hours			Overtime			Normal Working hours			Overtime		
	P1	P2	P3	P1	P2	P3	P1	P2	P3	P1	P2	P3
Machine 1	2	3	4	3	4	5	3	4	5	4	5	6
Machine 2	4	3	2	5	4	3	5	4	3	6	5	4
Machine 3	1	–	–	2	–	–	2	–	–	3	–	–

P1 = Nuts P2 = Bolts P3 = Washers

TABLE A.3.3. Table of additional data (TABD)

		SUMMER PERIOD			WINTER PERIOD		
		Nuts	Bolts	Washers	Nuts	Bolts	Washers
Sale price		10	10	9	11	11	10
Minimum demand		25	30	30	30	25	25
Storage data	Capacity	20	20	—			
	Cost	1	1	1			
	Resale value				2	2	1

TABLE A.3.4. Production schedule for two periods set out against demand

	Summer period			Winter period		
	Nuts	Bolts	Washers	Nuts	Bolts	Washers
Machine 1	xx.x	xx.x	xx.x	xx.x	xx.x	xx.x
Machine 2	xx.x	xx.x	xx.x	xx.x	xx.x	xx.x
Machine 3	xx.x	xx.x	xx.x	xx.x	xx.x	xx.x
Demand	xx.x	xx.x	xx.x	xx.x	xx.x	xx.x
Total	xx.x	xx.x	xx.x	xx.x	xx.x	xx.x

N.B. All production in 1000 lb weight of iron.

```
BASIS          STB
 XL  T1NP1M1    T1P1ST
 XL  T1NP2M1    T1P2ST
 XL  T1NP3M1    T1P3ST
 XL  T2NP1M1    T2P1ST
 XL  T2NP2M1    T2P2ST
 XL  T2NP3M1    T2P3ST
 LL  T1P1D
 LL  T1P2D
 LL  T1P3D
 LL  T2P1D
 LL  T2P2D
 LL  T2P3D
ENDATA
```

FIG. A.3.4.

two time periods, say WINTER and SUMMER. Any product left after the second time period has very little resale value. The necessary information concerning the operation of the company is set out in the tables TABH, TABD and TABC overleaf.

It is necessary to find an LP formulation which maximizes the profit of the company's operation over the two periods. The management also wishes to have a report of the optimum plan in a layout shown in Table A.3.4.

A.3.3b. A mathematical formulation

Let the four indices i, j, k, l be defined as

$i = 1, 2$ the index for the two time periods, summer and winter,
$j = 1, 2$ the index for the two modes of production, normal, overtime,
$k = 1, 2, 3$ the index for the three product types, P1, P2, P3,
$l = 1, 2, 3$ the index for the three machines, M1, M2, M3.

The following information relating to the problem are available:

In the table TABH,

t_{ijkl} number of hours required to produce one unit of the product type k on the machine l, in the time period i, using normal or overtime production j,
a_{ijl} machine availability in hours for the machine l in period i and mode j.

In the table TABD,

p_{ik} selling price, d_{ik} demand, } for the product type k in the time period i,
s_k storage cost for the product type k in one time period,
h_k the corresponding storage capacity,
r_k the final resale value value at the end.

In the table TABC,

c_{ijkl} the production cost in the category i, j, k, l.

The variables of the problem. Let,

x_{ijkl} denote the quantity to be produced in the category i, j, k, l,
y_{ik} denote the quantity of product k stored in the period i,
z_{ik} denote the quantity of product k sold in the period i.

The profit function of the problem may be expressed as

$$\text{Profit} = \sum_{i=1}^{2}\sum_{j=1}^{2}\sum_{k=1}^{3}\sum_{l=1}^{3}(p_{ik} - c_{ijkl})x_{ijkl}$$

$$- \sum_{k=1}^{3} s_k y_{1k} + \sum_{k=1}^{3}(r_k - p_{2k})y_{2k}½.$$

In an optimal plan Profit must be maximized subject to the constraints,

(i) machine availability,

$$\sum_{k=1}^{3} t_{ijkl} \cdot x_{ijkl} \leqslant a_{ijl}, \text{ for all } i, j, l,$$

(ii) stock balance in the two periods,

$$\sum_{j=1}^{2}\sum_{l=1}^{3} x_{1jkl} - y_{1k} - z_{1k} = 0, \text{ for period 1, and all } k,$$

TABLE A.3.5.

Variables → / Constraints ↓	T1NP1M1	T1NP1M2	T1NP1M3	T1NP2M1	T1NP2M2	T1NP3M1	T1NP3M2	T1OP1M1	T1OP1M2	T1OP1M3	T1OP2M1	T1OP2M2	T1OP3M1	T1OP3M2
PROFIT	8	6	9	7	7	5	7	7	5	8	6	6	4	6
T1M1AN	4			5		6								
T1M2AN		7			6		6							
T1M3AN			3											
T1M1AO								3			4		5	
T1M2AO									6			5		5
T1M3AO										2				
T2M1AN														
T2M2AN														
T2M3AN														
T2M1AO														
T2M2AO														
T2M3AO														
T1P1ST	1	1	1					1	1	1				
T1P2ST				1	1						1	1		
T1P3ST						1	1						1	1
T2P1ST														
T2P2ST														
T2P3ST														
BOUND LIM														
UP														
LO														

Variables → / Constraints ↓	T1P1STR	T1P2STR	T1P3STR
PROFIT	−1	−1	−1
T1M1AN			
T1M2AN			
T1M3AN			
T1M1AO			
T1M2AO			
T1M3AO			
T2M1AN			
T2M2AN			
T2M3AN			
T2M1AO			
T2M2AO			
T2M3AO			
T1P1ST	−1		
T1P2ST		−1	
T1P3ST			−1
T2P1ST	1		
T2P2ST		1	
T2P3ST			1
BOUND LIM			
UP	20		20
LO			

Variables → / Constraints ↓	T2NP1M1	T2NP1M2	T2NP1M3	T2NP2M1	T2NP2M2	T2NP3M1	T2NP3M2	T2OP1M1	T2OP1M2	T2OP1M3	T2OP2M1	T2OP2M2	T2OP3M1	T2OP3M2	T2P1STR	T2P2STR	T2P3STR
PROFIT	7	5	8	6	6	4	6	6	4	7	5	5	3	5	−9	−9	−9
T1M1AN																	
T1M2AN																	
T1M3AN																	
T1M1AO																	
T1M2AO																	
T1M3AO																	
T2M1AN	5			6		7											
T2M2AN		8			7		7										
T2M3AN			4														
T2M1AO								4			5			5			
T2M2AO									7			6		6			
T2M3AO										3							
T1P1ST																	
T1P2ST																	
T1P3ST																	
T2P1ST	1	1	1					1	1	1					−1		
T2P2ST				1	1						1	1				−1	
T2P3ST						1	1						1	1			−1
BOUND LIM																	
UP																	
LO																	

Variables → / Constraints ↓	T1P1D	T1P2D	T1P3D	T2P1D	T2P2D	T2P3D		RHS
PROFIT							N	
T1M1AN							LE	100
T1M2AN							LE	100
T1M3AN							LE	40
T1M1AO							LE	80
T1M2AO							LE	90
T1M3AO							LE	30
T2M1AN							LE	110
T2M2AN							LE	110
T2M3AN							LE	50
T2M1AO							LE	90
T2M2AO							LE	100
T2M3AO							LE	40
T1P1ST	−1						EQ	
T1P2ST		−1					EQ	
T1P3ST			−1				EQ	
T2P1ST				−1			EQ	
T2P2ST					−1		EQ	
T2P3ST						−1	EQ	
BOUND LIM								
UP								
LO	25	30	30	30	25	25		

and

$$\sum_{j=1}^{2} \sum_{l=1}^{3} x_{2jkl} + y_{1k} - y_{2k} - z_{2k} = 0, \text{ for period 2, and all } k$$

(iii) minimum demand to be satisfied

$$z_{ik} \geqslant d_{ik}, \text{ for all } i, \text{ and } k,$$

(iv) upper bound on storage,

$$y_{1k} \leqslant h_k, \text{ for all } k,$$

(v) non-negativity of the variables,

$$y_{ik} \geqslant 0, \text{ for all } i, k, \text{ and}$$

$$x_{ijkl} \geqslant 0, \text{ for all } i, j, k, l.$$

A.3.3c. Formulation of the problem for an LP user

To formulate the model in a form that can be processed by an OPTIMIZER it is necessary to construct names for the rows and columns of the model, specify the values of the matrix, cost, rhs coefficients and prepare these in a card image format. In the subsequent discussion input format described in Appendix 2 is used. It is noted there are three groups of variables in the model. These are x, y, z and let these groups be called PRODN, STORE, DEMAND

Naming the variables:

PRODN set variables; let these be called T1NP1M1 for x_{1111}, T1OP1M1 for x_{1211} etc.

STORE set variables; let these be called T1P1STR for y_{11}, etc.

DEMAND set variables; let these be called T1P1D for z_{11}, etc.,

It follows from the table entries that not all x_{ijkl}, hence, the corresponding Ti... M1 variables are defined.

Naming the constraints:

PROFIT the name of the objective row,

T1M1AN availability constraint cf(i) of machine M1 in time period T1 under Normal shift etc.,

T1P1ST stock balance equation cf(ii) in time period T1 for the product P1 etc.

Let the RHS column be called RHS, and the BOUNDS be called LIM. The corresponding LP user formulation of the model is derived and set out in Table A.3.5, also note that a partial starting basis STB can be suggested as shown in Fig. A.3.4.

A.3.4 SOME RAMIFICATIONS OF THE DATA STRUCTURE FACILITY, DATA DEFINITION AND DATA ENTRY

It follows from a cursory look at the tables TABH, TABD, TABC used in the last section that some structuring is already invoked in the rows (DOWN) and columns (ACROSS) of these tables. For instance in the table TABH (Table A.2.1) a set TH is made up of H1, H2, the production information in period 1 and period 2. In each time period the operation can be of the two types normal (N) or overtime (O); one is also concerned with the availability (AV) of the machines in these two time periods.

Finally, each of the normal (N) and overtime (O) operation is again broken down by product type, P1, P2, P3. It would be relatively easy to formulate the model if the following sets could be defined, viz. TH = {H1, H2}, H1 = {N, O, AV}, H2 = {N, O, AV}

$$N = \{\text{P1}, \text{P2}, \text{P3}\}, \qquad O = \{\text{P1}, \text{P2}, \text{P3}\} \ldots, \text{etc.},$$

The structure facility of UIMP achieves exactly this: it allows such elements as TH, H1, N, P1 etc. to be defined and used in a meaningful way. In the opinion of the author if the input tables and the data are properly structured then this already leads a longway into the compact yet comprehensive definition of the model.

To generate the model already formulated in Section 3 it is necessary to write an input (program) module in UIMP. The module is called MGTABS and is set out overleaf. Various statements in the module as identified by their sequence numbers are annotated.

```
                                                C
LABEL C              INSTRUCTIONS               O
      O                                         L
      L                                         7  SEQ. NO
      6                                         3
      ↓                                         ↓
        MODULE  MGTABS                          1

        LET TH BE H1, H2/H1, H2 BE N, O, AV/N,
      — O, AV/N, O BE P1, P2, P3/AV BE AN, AO/
      — MACH BE M1, M2, M3                      2

        LET TD BE T1, T2/T1, T2 BE P1, P2, P3/
      — DROW BE PRICE, DEM, STR/STR BE CAP,
      — COST, RESL                              3

        LET TC BE C1, C2/C1, C2 BE N, O         4
```

```
     LET TABLE TABH BE MACH DOWN BY
 —                 TH ACROSS
                   TABD BE DROW DOWN BY
 —                 TD ACROSS
                   TABD BE MACH DOWN BY
 —                 TC ACROSS                          5
     INPUT DATA BY ROW                                6

     TABH/4, 5, 6, 3, 4, 5, 100, 80, 5, 6, 7, 4, 5, 6,
 —             110, 90                                7
               7, 6, 6, 6, 5, 5, 100, 90, 8, 7, 7, 7, 6, 6,
 —             110, 100                               8
               3, , , 2, , , 40, 30, 4, , , 3, , , 50, 40   9

     TABD/10, 10, 9, 11, 11, 10                      10
               25, 30, 30, 30, 25, 25                11
               20, 20                                12
               1, 1, 1                               13
               , , , 2, 2, 1                         14

     TABC/2, 3, 4, 3, 4, 5, 3, 4, 5, 4, 5, 6         15
               4, 3, 2, 5, 4, 3, 5, 4, 3, 6, 5, 4    16
               1, , , 2, , , 2, , , 3                17
 ENDATA                                              18
 ENDMGTABS                                           19
```

Annotation of the module MGTABS

Sequence No. 2. This statement declares the structure required to represent the columns of the table TABH; Conceptually this has a tree structure, Fig. A.3.5.

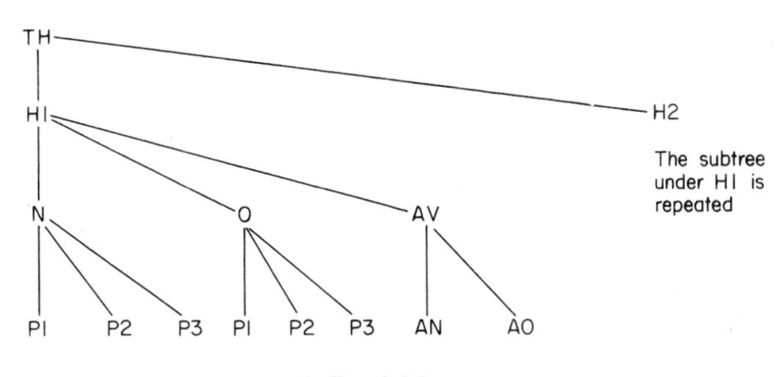

FIG. A.3.5.

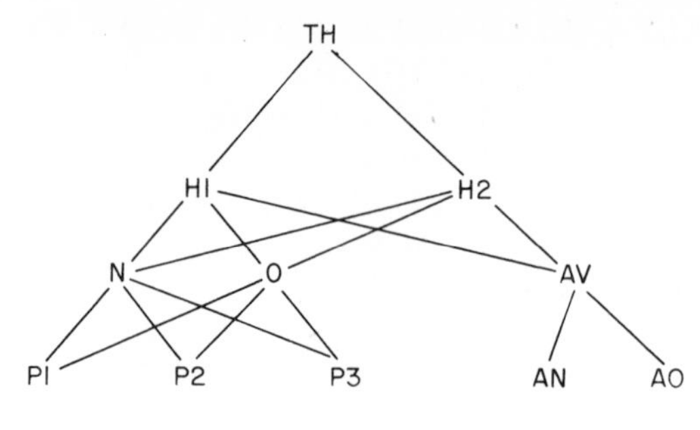

FIG. A.3.6.

However, because of the repetition of the subtree of H2, the network shown in Fig. A.3.6 is a more convenient representation of this structure.

All the structures used in the program are set out in the next section in this network form, these diagrams are useful to understand the structure facility and its use in the program.

No 3, 4 declare all the structures necessary for the tables, TABH, TABC, TABD.
No 5 declares the structure of the data tables.
No 6 declares a data input mode in which the data tables can be filled in.
No 7–17 the data tables are input in free format.
No 18 signals the end of data input.
No 19 signals the end of the module MGTABS.

It is natural to define implicitly the structures of the rows and the columns of the tables at the time of preparing the input data tables. Therefore a display format specification of the structure by using the input data tables is also possible in the UIMP system. The program module FORMTABS below illustrates this alternative way of carrying out the task of MGTABS module.

```
                           INSTRUCTIONS                  C
LABEL C                                                  O
      O                                                  L  SEQ NO
      L                                                  7
      6                                                  3
      ↓                                                  ↓
        MODULE FORMTABS
        INPUT TABLE
```

```
TABH/TH
        H1                              ,
        N        ;     O     , AV       ;
        P1, P2, P3; P1, P2, P3; AN, AO;
            H2                  ;
            N    ,     O     , AV;
        P1, P2, P3; P1, P2, P3; AN, AO;

MACH M1 = 4, 5, 6, 3, 4, 5, 100, 80, 5, 6, 7, 4, 5,
          6, 110, 90
     M2 = 7, 6, 6, 5, 5, 100, 90, 8, 7, 7, 7, 6, 6,
          110, 100
     M3 = 3, , , 2, , , 40, 30, 4, , , 3, , , 50, 40
ENDTABH
TABC/TC
     .
     .        .  .  .
     .
ENDTABC
TABD/TD
     .
     .        .  .  .
     .
ENDTABD
ENDTABLE
ENDFORMTABS
```

A.3.5 THE ILLUSTRATIVE EXAMPLE FULLY CODED AND ANNOTATED

To complete the description of the MG and RW programs written in UIMP it is necessary to describe the four separate program modules which are used for this task. These modules are,

MGTABS
MGGEN
RWTABS
REPORT.

The module MGTABS is already annotated in Section 4. In this section the three remaining modules are set out and their steps are annotated.

INSTRUCTIONS

```
                                                         C
LABEL C                                                  O SEQ NO
      O                                                  L
      L                                                  7
      6                                                  3
      ↓                                                  ↓
        MODULE MGGEN                                     1

        GET MGTABS                                       2

        LET TA BE A1, A2                                 3

        LET A1, A2, BE AN, AO                            4

        LET KP, JW, IT, LM, JA, IA, IC BE
      — STRUCTURE VARIABLE                               5

        LET PROD BE VARIABLE CONSTRUCTED
      — LM IN MACH BY KP IN JW IN IT IN TC
      —   CODE (TD(I: IT)) & (JW) & (KP) & (LM)          6
        LET DEMAND BE VARIABLE
      —   CONSTRUCTED KP IN IT IN TD CODE
      —   (IT) & (KP) & "D"                              7
        LET STORE BE VARIABLE CONSTRUCTED
      —   KP IN IT IN TD CODE (IT) & (KP) & "STR"        8
        LET AVAIL BE CONSTRAINT TYPE LE
      —   CONSTRUCTED LM IN MACH BY JW IN
      —     IT IN TA CODE (TD(I: IT)) & (LM) & (JW)      9
        LET BAL BE CONSTRAINT TYPE EQ
      —   CONSTRUCTED KP IN IT IN TD CODE
      —   (IT) & (KP) & "ST"                             10
        LET OBJ BE OBJECTIVE CODE "PROFIT"               11
        ROW OBJ IS SUM KP IN JW IN IT IN TC
      —   SUM LM IN MACH (TABD(PRICE, TD (I:
      —   IT) (I: KP))—TABC (LM, KP)) MPLYING
      —   PROD (LM, KP)                                  12
      —   MINUS SUM KP IN TI IN TD TABD
      —   (COST IN STR, KP) MPLYING STORE
      —   (KP PLUS SUM KP IN TS IN TD
```

```
—      (TABD(RESL IN STR, KP)—TABD(PRICE,
—      KP)) MPLYING STORE (KP)
    FOR JA IN IA IN TA DO BEGIN                     13
       IT:=TH(I: IA); JW:=IT(I: JA); IC:=(TC(I:
—         IA)                                          14
       FOR LM IN MACH DO                               15
—         ROW AVAIL(LM, JA) IS SUM KP IN JW
—         TABH (LM, KP) MPLYING PROD(LM,
—          KP IN JW IN IC)
          RHS "RHS" VALUE TABH (LM, JA IN
—         AV IN IT)
    END                                                16
```

INSTRUCTIONS

```
                                                      C
LABEL  C                                              O  SEQ NO
       O                                              L
       L                                              7
       6                                              3
       ↓                                              ↓
    FOR KP IN T1 IN TD DO                             17
       ROW BAL(KP) IS SUM LM IN MACH
—      SUM JW IN C1 IN TC
—      PROD(LM, KP IN JW) MINUS STORE
—      (KP) MINUS DEMAND(KP)
—      IN BASIS "STB" VARIABLE PROD(M1,
—      KP IN N IN C1) PIVOTS  OUT  ROW
    FOR KP IN T2 IN TD DO                             18
       ROW BAL(KP) IS SUM LM IN MACH
—      SUM JW IN C2 IN TC
—      PROD(LM, KP IN JW) MINUS STORE
—      (KP) MINUS DEMAND(KP)
—      PLUS STORE(KP IN T1)
—      IN BASIS "STB" VARIABLE PROD(M1,
—      KP IN N IN C2) PIVOTS OUT ROW
    FOR IN T1 IN TD DO                                19
       COLUMN STORE (KP) BOUNDS "LIM"
—      UP VALUE TABD(CAP IN STR, KP)
    FOR KP IN IT IN TD DO                             20
```

```
—    COLUMN DEMAND (KP) BOUNDS "LIM"
—    LO VALUE TABD(DEM, KP)
    END MGGEN                                   21
```

Annotation of the module MGGEN

No 2 This statement extracts from the MP data-base the compiled module, so that the structures, tables etc can be accessed.

No 3, 4, 5 declare structures and structure variables.

No 6. The group of variables called PROD is declared this corresponds to the x_{ijkl} group of variables (see 3c). The indices l, (ijk) are defined by LM IN MACH and KP IN JW IN IT IN TC. The clause CODE introduces the name expression for constructing individual variable names which are T1NP1M1, T1NP1M2, T1NP1M3, T1NP2M1 . . . etc.

No 7,... 10 declare groups of variables and constraints and the objective(s).

No 12 generates the objective row 'PROFIT' which is equivalent to the form

$$\sum_i \sum_j \sum_k \sum_l (p_{ik} - c_{ijkl})x_{ijkl} - \sum_k s_k y_{1k} + \sum_k (r_k - p_{2k})y_{2k}$$

The table position references such as TABD(PRICE, TD(I: IT) (I: KP)) contain element context expressions—in this case PRICE and TD(I: IT) (I: KP). Note that since the first structure for TABD is DROW, "IN DROW" is implicit.

Thus the above can be re-expressed as TABD(PRICE IN DROW, TD(I: IT) (I: KP) IN TD(I: IT) IN TD).

No 13, . 16. These generate the constraints

$$\sum_k t_{ijkl} \cdot x_{ijkl} \leqslant a_{ijl}, \quad \text{for all, } i, j, l.$$

Note that one is free to use any element context expression as a subscript without causing reassignment of structure variables. Thus on line 3 of the statement the subscript 'KP IN JW IN IC' does not in any way change the fact that KP and JW are assigned in the context "KP IN JW IN IT IN TH".

No 17, 18 set up the starting basis 'STB' for use by the LP OPTIMIZER.

No 19, 20. Bound values are specified by column generator clauses.

```
                          INSTRUCTIONS                    C
LABEL C                                                   O
      O                                                   L
      L                                                   7
      6                                                   3
      ↓                                                   ↓
        MODULE RWTABS                                     1
        GET MGTABS                                        2
        LET RTROW BE MACH, DEM, TOT                       3
        LET TABLE RTAB BE RTROW DOWN BY
    —     TD ACROSS                                       4
        LET I, K, R, L BE STRUCTURE VARIABLE
        INPUT TEXT                                        5
        I IN TD        ,          ,           ,           6
        * (1 BL LINE)                                     7
                       SUMMER PERIOD
                                  WINTER PERIOD           8
        K IN I       ,     ,     ,     ,    ,    ,        9
        * (1 BL LINE)                                     10
                       NUTS BOLTS WASHERS
                       NUTS BOLTS WASHERS                 11
        R IN RTROW ,–,           ,            ,           12
        * (2 BLANK                                        13
        *    LINES)                                       14
                   S   DEMAND   TOTAL                     15
        L IN MACH ,           ,          ,                16
                      MACHINE 1 MACHINE 2
                      MACHINE 3                           17
        END TEXT                                          18
        END RWTABS                                        19
```

ANNOTATION OF MODULE RWTABS

No 3, 4 These statements set up a table RTAB in the Layout required for the report.

No 5 Enters a Heading-Text input mode that assigns text to structural entities.

No 6 To read in texts for entities T1 and T2 in the structure of TD. A field-width is supplied for each entity by the tab markers (,) and any text that is too long for the field width is automatically “folded” onto extra lines when printing takes place. Thus the print-layout can be controlled both vertically and horizontally.

No 7 Blank lines are automatically put on the print-out. Note that the space in front of the first tab-marker can be used for commentary.

No 12 ',–,' indicates that the field-width for entity MACH is zero.

Diagram of all column-wise structures

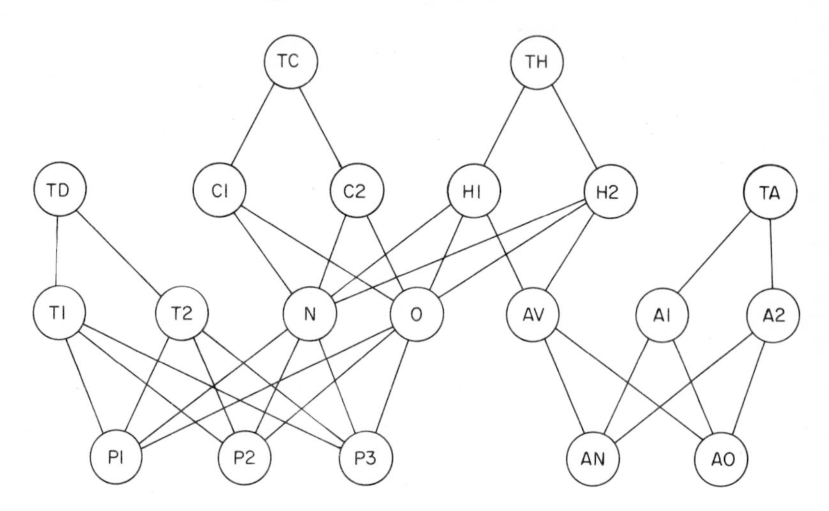

Diagram of all row-wise structures

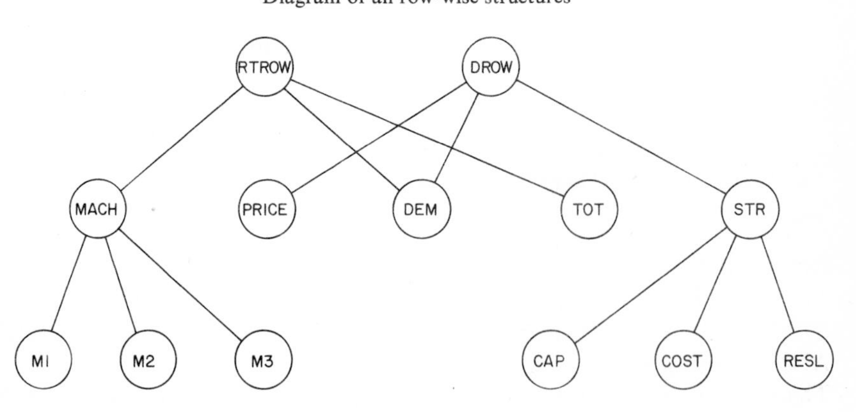

FIG. A.3.7.

```
                         INSTRUCTIONS                    C
       C                                                 O  SEQ NO
LABEL  O                                                 L
       L                                                 7
       6                                                 3
       ↓                                                 ↓
         MODULE REPORT                                   1
         GET RWTABS, MGGEN                               2
         LET F1 BE FORMAT (13X, 'PRODUCTION
       —                     SCHEDULE FOR TWO
       —                     PERIODS'                    3
       —                     13X, '————————'
       —                     22X, 'SET OUT
       —                     AGAINST DEMANDS'
       —                     22X, '————————'
         LET F2 BE FORMAT (////                          4
       —  "NB. ALL PRODUCTION IN 1000 LBS
       —     WEIGHT OF IRON")
         FOR KP IN IT IN TD DO BEGIN RTAB
                                (TOT, KP):=O . O         5
                                RTAB(DEM, KP):=
       —                        TABD(DEM, KP)
                                END                      6
         FOR LM IN MACH DO FOR KP IN IT IN
       —     TD DO BEGIN                                 7
             IT:=TC(I: IT)                               8
             RTAB(LM, KP):=X: PROD(LM, KP IN
       —                    N IN IC)                     9
       —                  + X: PROD(LM, KP IN
       —                    O IN IC)
             RTAB(TOT, KP):=RTAB(TOT, KP) +
       —       RTAB(LM, KP) END                          10
         PRINTABLE RTAB HEADER F1 TRAILER
       —   F2                                            11
         END REPORT                                      12
```

ANNOTATION OF MODULE REPORT

This module prints the actual report

No 3, 4 Declare heading and trailer formats. Format text interpreted as in FORTRAN is given within brackets.

No 5, 6 Zeros the total line and copies demand figures from the table TABD.

No 9 Enters solution values (denoted by "X:") in rows MACH of RTAB. Normal and overtime are added.

No 10 Accumulate the total.

No 11 Prints the final result altogether in one go.

REFERENCES

1. A Brief Survey of the Currently Available Matrix Generator Report Writer Systems, Internal Report, UNICOM Consultants Ltd, 1974.
2. Backus, J. W., The Syntax and Semantics of the proposed International Algebraic Language of the Zurich, ACM-GAMM-Conference, ICIP, PARIS, 1959.
3. User Manual for UIMP, UNICOM Consultants Ltd, U.K., 1975.
4. Mathematical Programming System Extended (MPSX), Program Number 5734 XM4, IBM Corporation, New York (1971).